GUAVA
(*Psidium guajava* L.)

The Authors

Ms. Sohnika Rani is doing Ph.D in the Division of Fruit Science, SKUAST-Jammu. She has done B.Sc Agriculture and M.Sc in Fruit Science from SKUAST-Jammu. She has published 06 articles.

Dr. Akash Sharma is working as Asstt. Professor in the Division of Fruit Science, SKUAST-J, Chatha-Jammu. He isassociatedwith teaching of UG & PG students from 2008. His research papers is published in high impact journals of national and international levels. He has 52 no. of publications in his credit (25 national, 05 International). Dr. Sharma is associated with six professional horticulture societies and is also member in the editorial team of the Journals.

Prof. V. K. Wali is presently working as Head, Division of Fruit Science, Faculty of Agriculture, SKUAST-Jammu. He has handled 12 institutional and 15 externally funded projects. He has published 95 articles in various national and international journals. He is a Life-member of 5 National Horticultural Societies of India.

GUAVA

(*Psidium guajava* L.)

Sohnika Rani
Akash Sharma
V. K. Wali

2018
Scholars World
A Division of
Astral International Pvt. Ltd.
New Delhi – 110 002

ISBN 9789389569940 (Int Edition)

Published by : **Scholars World**
A Division of
Astral International Pvt. Ltd.
– ISO 9001:2015 Certified Company –
4736/23, Ansari Road, Darya Ganj
New Delhi-110 002
Ph. 011-4354 9197, 2327 8134
E-mail: info@astralint.com
Website: www.astralint.com

Digitally Printed at : **Replika Press Pvt. Ltd.**

Preface

The present investigation entitled "Propagation studies in guava (*Psidium guajava* L.) cv.L-49 under Jammu sub-tropics" was carried out at Udheywalla, Faculty of Agriculture, Sher-e-Kashmir University of Agricultural Sciences and Technology of Jammu.

Propagation methods performed in the experiment were cutting, air layering, patch budding, wedge grafting in the month of July, August and September during 15th to 21st respectively. It was observed that among all propagation methods, patch budding performed during 15th to 21st of August showed highest per cent success (92.07 per cent) whereas, patch budding (87.61 per cent) also showed highest per cent success in the month of September which was at par with wedge grafting (86.57 per cent) after 90 days of propagation. After 180 days of guava propagation, maximum number of new shoots per plant (8.97), number of leaves per plant (22.39), scion thickness (1.34 cm) were recorded in wedge grafting performed during 15th to 21st of September while, air layering during 15th to 21st of August took minimum days to sprout (11.75) and maximum increase in plant height (61.89 cm).

The results in the II experiment indicated that among different soil media used for planting guava softwood cuttings, vermiculite + sand + FYM (1:1:1) mixture applied showed maximum rooted cuttings (78.69 per cent), number of new shoots per plant (10.42), plant height (26.49 cm), number of leaves per plant (25.08) and stem thickness (1.14 cm) while, cuttings planted in perlite + sand + FYM (1:1:1) took minimum days to sprout (8.95) during 15th to 21st of August.

Therefore it can be concluded that patch budding performed during 15th to 21st of August was found to be the best method for guava propagation under Jammu sub-tropics.

In the second experiment it may be concluded that guava cuttings planted in soil media of Vermiculite + sand +FYM (1:1:1) during 15th to 21st of August was found to be best suitable media for guava propagation under Jammu sub-tropics.

Sohnika Rani

Akash Sharma

V.K. Wali

Contents

List of Figures

List of Tables

Chapter 1

Introduction

Guava (*Psidium guajava* L.) belonging to family Myrtaceae is originated in tropical South America (Pathak and Ojha 1993). It is considered to be one of the exquisite, nutritionally valuable and remunerative crop. Besides its high nutritional value, it bears heavy crop every year and gives good economic returns (Singh *et al.*, 2007). The tree is fairly salt and drought-resistant and can be grown on a variety of soils. This has prompted several farmers to take up guava orcharding on a commercial scale (Singh and Bajpai, 2003). In recent years, guava is getting popularity in the international trade due to its nutritional value and processed products (Singh, 2005). Guava exceeds most other fruit trees in productivity, hardiness and adaptability. The fruits are used for both fresh consumption and processing, it has great market potential due to their delicious taste and aroma. The fruits after removal of seeds may be utilized to make products such as jam, jelly, cheese, juice, canned segments, nectar etc. They are used mostly for squashes and juices, however the most commercial use of guava is for jelly preparation due to good source of pectin (Adsule and Kadam, 1995).

In view of the high return and potential for processing there is tremendous scope for bringing substantial additional area under guava crop in India as well as with the changing horticultural scenario in India, demand of genuine planting material has increased tremendously. The area under guava cultivation in India is estimated to be 322 thousand hectare with an annual production of 2510 MT (Anonymous, 2012). Demand of planting material has encouraged establishment of a large number of nurseries to cater the acute shortage of quality planting material. Non availability of quality planting material and consequent substitution of poor quality seedling has adversely affected the guava production and productivity (Singh *et al.*, 2005). Therefore, rapid and successful propagational technique is required to cater the planting material of guava throughout the year. While, choosing a particular technique for propagation of guava, the time of operation and method should be taken into consideration. As the success of each method vary from region to region due to variation in agro climatic conditions. Any particular method which may be successful at one place may not prove useful at other. Similarly, a particular method successfully adopted will vary from place to place due to environmental factors such as temperature, relative humidity etc.

In Jammu and Kashmir guava culture has shown tremendous potential owing to be the most remunerative crop for last one decade. Despite of erratic climatic condition/drought like situations which in turn has boosted area under this crop. The area under guava cultivation in Jammu is estimated to be 2372 hectare with an annual production of 5692 MT (Anonymous, 2012). The Jammu region represent distinct agro climatic conditions characterized by high humidity, hot summer and cold winter. Therefore, for meeting the demand for true to type plants of guava and its rapid and successful propagation through vegetative propagation, it is essential that various practices for vegetative propagation of guava be evolved and evaluated for this region.

Guava is propagated by many methods of propagation and different degree of success have been achieved by various propagation techniques *viz.* , 93.3 per cent with patch budding (Kumar *et al.*, 2007), 77.33 per cent with wedge grafting (Gurjar *et al.*,2012), 71.22 per cent with cuttings (Rahman *et al.*, 2003) and 83.15 per cent rooting success with air layering (Rymbai and Reddy, 2010). Though various methods for propagation of guava are widely used but still many factors affect the propagation in different regions such as in propagation through seeds, the seeds are sown immediately after extraction, seedlings to tree is generally long lived but are not true to type, which bear fruit of variable size, variation occurs not only with respect to the productivity of the tree, but also in various quality parameters *viz.* , size of fruits, content of total soluble solids, sugars and above all vitamin C content. Seed originated plants cannot maintain genetic purity of the variety due to the segregation recombination of characters during sexual reproduction. Besides, the seedlings have a long juvenile phase. Guava is difficult to root, even if cuttings are taken from vigorously growing new shoots (Khattak, *et al.*, 1983). Propagation by cutting, grafting and patch budding have been attempted by various investigators (Samson, 1986).

Soil media is considered an integral part of propagation (Loach, 1988) and percentage rooting and quality of roots produced are directly influenced by the medium. Preparation of suitable soil media includes the standardization of texture and nutrient status. Perlite is far the most used rooting substrate in olive producing countries. Mixtures such as perlite plus peat, coconut fiber or vermiculite have also given good results (Fabbri *et al.*, 2004 and Sutter, 2005). A suitable selection of soil media components and its ratio will decide the success of the planting medium. In present study combinations of various soil media were used for planting guava cuttings to find out the best soil media for guava propagation.

For better survival and establishment in the field and to achieve good plant population from vegetative method for meeting the growing demand of guava planting material in Jammu sub-tropics, standardization of propagation method, time of propagation and soil media for obtaining maximum success is very important as recommendation made in other region cannot be out rightly applied under the agro climatic conditions of Jammu. Keeping in view the problem in method and time of propagation, the present studies were undertaken with following objective:

- ☆ To standardize most suitable method for guava propagation under Jammu sub-tropics.
- ☆ To find out best soil media and time for guava propagation through cuttings under Jammu sub-tropics.

Chapter 2
Review

The greatest handicap in guava plantation is discriminate multiplication and non availability of quality planting material that adversely affect the guava production and productivity. The initial planting material is basic requirement on which the final crop depends both in quality and quantity. In view of the high return and the potential for processing, there is a tremendous scope for bringing substantial additional area under guava crop in India. Therefore, rapid methods of propagation become very important when planting materials are limited due to scarcity of clone or varieties or due to sudden expansion in acreage. Thus, it has become imperative to standardize the time and propagation methods for guava under open conditions. A comprehensive review of work done in India and abroad on propagation studies in guava (*Psidium guajava* L.) are presented below under suitable headings.

2.1 Influence of Time and Propagation Methods on Per cent Success

Pandey *et al.* (1979) reported that swollen buds gave better bud take than dormant buds and patch-budding showed highest (90 per cent) bud take in May as compared to chip budding performed during April to August in guava. Similarly, Mehrotra and Gupta (1984) revealed that the highest (70.12 per cent) success in patch budding of guava was obtained when seedlings were budded during May which was at par with June (68.75 per cent) and July patch budding (60.00 per cent). The patch budding success during May, June and July was at par with each other but significantly superior over April (43.75 per cent), August (42.50 per cent) and September months. Results of Rao *et al.* (1984) revealed that July and August months were optimum time for budding of guava where maximum success of 74 per cent in patch budding during July and 62 per cent in August, with forket budding were obtained. Similarly, Gupta and Mehrotra (1985) obtained highest (82.50 per cent) patch budding success in May which was closely followed by 80.00 per cent success in June where, the patch budding success in May and June was significantly better than that from April to September.

Kilany and Gabr (1986) reported that rooting in hardwood cuttings of guava was very poor (1.67-4.67 per cent only), it ranged from 18.3 to 57.5 per cent in leaf-bud cuttings and was highest (81.4 per cent) in semi-hardwood cuttings treated with 2500 ppm IBA + 10 ppm α -naphthol planted during September in a 1:1 mixture of sand and peat moss. Kaundal *et al.* (1987) obtained maximum success (87.50 per cent) in patch budding during May as compared to shield budding performed at monthly intervals from April to September. Rahman *et al.* (1988) revealed that leafy tip cuttings of guava gave 90.11 per cent, 94.44 per cent and 94.44 per cent rooting at 3, 6 and 12 ppm paclobutrazol, respectively, after six weeks of planting in July-August whereas, no rooting was observed in control.

Rahman *et al.* (1991) reported that 10-12 cm long guava tip cuttings with at least 4 leaves, dipped in paclobutrazol for 24 hours and planted between mid June and end of September in sand under unheated greenhouse with 78-80 per cent RH and natural light conditions gave the highest (94 per cent) rooting percentage in mid-August planting. Sharma *et al.* (1991) revealed that air layering carried out on 10th July resulted in the highest (67.70 per cent) per cent success as compared to air layering performed on 10th July, 25th July or 10th August in 1990 on eight year-old guava tree. Singh *et al.* (1992) reported highest survival (75.63 per cent) with 75 per cent defoliation followed by complete defoliation (64.18 per cent) in air layers of guava at 45 and 120 days after planting in the field.

Results of Aulakh (1998) showed highest (95.60 per cent) percentage of successful plants in guava when patch budding was performed on 14th June which was closely followed by patch budding performed on 29th June whereas, minimum percentage of successful plants (25 per cent) were recorded on 29th July which was followed by 15th May. Singh and Pandey (1998) reported that the best time for budding in guava was July followed by August and poor success was observed in February, March, April, May and June. Bhagat *et al.* (1999) reported highest rooting (94.67 per cent) and survival (78.33 per cent) in air layers of guava when treated with 4500 ppm IBA.

Manan *et al.* (2002) reported highest per cent success (51.24 per cent) in guava cuttings treated with IBA at 500 ppm. Rahman *et al.* (2003) concluded that maximum (71.22 per cent) sprouting per cutting within 17.68 days in softwood cuttings treated with 1000 ppm NAA in the month of August. Results of Ayaz *et al.* (2004) revealed that treatment of guava cuttings with 60 ppm paclobutrazol resulted in maximum (73.3 per cent) cutting success in fresh softwood cuttings of guava having 3-4 leaves. Patil (2004) reported that maximum success was achieved through wedge grafting (91.6 per cent) in mid- August which was followed by patch budding (85.0 per cent) among all propagation methods and mid-August wedge grafting took minimum days to sprout with maximum success which was followed by grafting in July. Ghosh and Ranjan (2005) reported that air layering in September, October and November resulted in highest (85 per cent) rooting success when performed on the 10th of each month from January 2001 to December 2002 in guava cv. L 49.

Results of Patel *et al.* (2005) revealed that in patch budding highest (84.08 per cent) success and minimum number of days (27.67) for sprouting in cultivar Allahabad Safeda was observed when budding was performed on seven guava cultivars in the month of February-March. Ullah *et al.* (2005) reported maximum (71.22 per cent) sprouting in softwood cuttings of guava cv. Allaabadi treated with paclobutrazol in August. The highest (60 per cent) rooting percentage was observed in the cuttings treated with 0.4 per cent IBA solution which was followed by rooting in cuttings treated with 0.2 per cent IBA (Abdullah *et al.*, 2006). Similarly, Kumar *et al.* (2007) reported maximum (93.30 per cent) budding success in mid June patch budding which was followed by end June and minimum number of days (18.0) were required for bud sprouting in mid June and maximum number of days (36.7) for budding done in the end of September in guava.

Results of Babu and Yadav (2007) showed that the response of guava to patch budding during third week of February, was found to be excellent with respect to time taken for bud-take (18 days), per cent bud sprout (95 per cent) and survival percentage (90.0 per cent) at two months after patch budding. Similarly, Patel *et al.* (2007) reported that highest (84.08 per cent) patch budding success and minimum (27.67) number of days taken for sprouting in guava cultivar Allahabad Safeda during February-March. Results of Singh *et al.* (2007) revealed that maximum success of wedge grafting was obtained in greenhouse (88.63-94.33 per cent) as well as in open field conditions (66.6-78.63 per cent) during November to February in guava (*Psidium guajava* L.) cultivars Allahabad Safeda and Sardar. Ahmad *et al.* (2007) reported that patch budding in walnut resulted in budding success of 22.90 per cent in poly house and 20.33 per cent in open field conditions while, 30 July was recorded as best time for maximum budding success in walnut. Babu *et al.* (2009) concluded that guava is highly amenable for patch budding and February third week is the most appropriate period for patch budding under subtropical conditions of Northeast India and revealed outstanding performance of Hybrid-1 with minimum duration for budtake (18.0 days) and maximum bud sprout (95.0 per cent). Marinho (2009) reported that fourty day after planting, 76 per cent of the minicuttings rooted and emitted aerial part and thirty-five days after been planted, these minicuttings, with average length of 13.56 mm, presented 100 per cent of rooting.

Results of Rymbai and Reddy (2010) revealed that high percentage of rooting and root characters of air layers of guava have been successfully achieved by exogenous application of IBA at 4000 ppm and air layering performed on 15th August gave maximum rooting success (77.94 per cent). Negi *et al.* (2010) reported that among all the methods of propagation maximum success (80.33 per cent) was recorded with patch budding in anola in the month of May. Visen *et al.* (2010) reported that wedge grafting has tremendous potential for multiplying plants rapidly either in greenhouse or open conditions in guava (*Psidium guajava*) and obtained maximum (81.71 per cent) success of wedge grafting in greenhouse during September and December. Mir and Kumar (2011) recorded maximum success (76.66 per cent) with wedge grafting among others methods of propagation under polyhouse conditions during 4 th week of February in walnut. Rymbai and Reddy (2011) reported that highest (77.94 per cent) rooting was obtained in 15th August air layering of guava and showed a possibility of obtaining good quality planting material using air layering (at 15th June, 15th July and 15th August). Singh *et al.* (2011) reported success in wedge grafting ranging from 28-99 per cent being maximum (99 per cent) in guava cultivars in the month of February in open field conditions.

Gurjar *et al.* (2012) reported that wedge grafting in anola resulted in maximum (93.62 per cent) bud brusting percentage under poly house conditions in the month of 1st March and (54.73 per cent) in open field. The maximum graft survival percentage was recorded in the month of 1st March in poly house condition (98.30 per cent) and open field condition (76 per cent). Results of Syamal *et al.* (2012) revealed that wedge grafting in the month of July gave better result in polyhouse (77.17 per cent) as well as in open field condition (66.43 per cent). Similarly, Gurjar *et al.* (2012) reported that wedge grafting in guava showed maximum bud bursting percentage in grafts in the month of Ist November in polyhouse and open field condition (67.65 per cent) in the month of 1st February. The maximum graft survival percentage was recorded in the month of 1st January in poly house (94.08 per cent) and open field condition (77.33 per cent) in the month of 15th January. Kareem *et al.* (2013) reported that softwood cuttings of guava treated with IBA 4000 ppm gave maximum (92.17 per cent) survival percentage of plants which was followed by (85.50 per cent) with IBA 2000 ppm in month of August.

2.2 Influence of Time and Propagation Methods on Growth Characteristics

Results of Mehrotra and Gupta (1984) recorded maximum (20.5 cm) sprout length with patch budding in June which was closely followed by May (20.3 cm), April (18.8 cm) and July (16.4 cm) budding which were significantly higher than August (7.9 cm) and September (3.7 cm) budding after one month of patch budding. Gupta and Mehrotra (1985) in their study of patch budding recorded that highest (22.7 cm) shoot length was obtained in May which was closely followed by June patch budding (22.5 cm) and shoot length in May and June were at par with each other but significantly superior to other months. Similarly, Kaundal *et al.* (1987) while comparing patch and shield method of budding recorded maximum (25.26 cm) budling length in May with patch budding which was significantly greater than the budling length in case of patch budding done in August (10.79 cm) and September (5.29 cm). Rahman *et al.* (1988) reported maximum (12.81cm) root length in semi hardwood and softwood guava cuttings when treated with 1000 ppm NAA in July – August. Results of Sharma *et al.* (1991) revealed that treatment of air layers with 10000 ppm IBA resulted in the highest number, length, diameter and weight of roots after carrying out air layering on 10 July, 25 July or 10 August in 1990 on eight-year-old guava tree. Patel and Pasaliya (1995) reported that NAA at 9000 ppm applied immediately after ringing gave the highest number of primary and secondary roots and heaviest root fresh and dry weights when IBA, NAA or IAA at 3000, 6000, 9000 ppm was applied after ringing or 10 to 20 days later on rooting in air-layered shoots of guava cv. Lucknow-49.

Aulakh (1998) reported that maximum (46.6 cm) shoot length was obtained on 14 June followed by 29 June patch budding and minimum (18.7 cm) shoot length was recorded on 29 July followed by 15 May (20.3 cm). Athani *et al.* (1999) reported that longest (11.15 cm) root length was noticed in cultivars GW-1 and GR-2, SR-1 has the shortest roots (2.8 cm), CIW-4 had the highest number of roots (12.3), and SR-3 had the lowest number of roots (1) when air layering was performed in guava cultivars. The highest (23.75) number of roots was recorded in the cuttings treated with IBA at 4000 ppm and significantly maximum (4.13 cm) root length was noted in the cuttings treated with IAA 3000 ppm in April (Wahab *et al.*, 2001). Similarly, Rahman *et al.* (2003) revealed that softwood cuttings treated with paclobutrazol gave good performance as compared to IBA and NAA in August. They obtained more branches (3.44), maximum root-weight (1.46 g) and more number of branches (3.44) in softwood cutting treated with paclobutrazol at the 100 ppm solution while, maximum (59.66) number of roots and lengthy shoot (8.24 cm) was recorded in softwood cuttings treated with IBA at 1000 ppm. Similarly, early sprouting (17.68 days) and maximum root-length of 12.81 cm was observed in softwood cutting, treated with NAA at concentration of 1000 ppm.

Results of Ayaz *et al.* (2004) revealed that 60 ppm paclobutrazol resulted in maximum rooting (69.5 per cent), shoot length (24.3 cm), number of branches (4.3), number of roots (87.1) and root volume per plant (1.64 cm^3) in fresh softwood cuttings of guava having 3-4 leaves. Among various dipping period five hours dipping resulted in maximum rooting (30.1 per cent), shoot length (10.6 cm) and number of roots (47.2), while four hours dipping resulted in the maximum number of branches (2.6) and root volume per plant (1.05 cm^3). Kumar and Syamal (2005) reported that IBA at 3000 ppm recorded the highest value for mean number (14.80) and length (11.30 cm) of primary roots per air layer, average number of secondary roots (10.72), while IBA at 4000 ppm recorded the highest value for diameter of roots (2.30 mm) when guava shoots of 1 to 2 years of age were air layered.

Patel *et al.* (2005) reported that maximum (36.89 cm) length of sprout, leaves/ plant (29.67) and leaf width (5.59 cm) were recorded in Allahabad Safeda, while maximum rootstock girth (3.82 cm), sprout girth (2.71 cm) and leaf length (10.95 cm) were recorded in cultivar hybrid-1 among air layered seven cultivars of guava. Ullah *et al.* (2005) observed more number of branches (3.44), maximum root weight (1.46 g) and better survival (57.22 per cent) in softwood cuttings of guava treated with paclobutrazol at 1000 ppm solution in August. Maximum number of roots (59.66) and lengthy shoot (8.24 cm) were recorded in soft wood cuttings of guava treated with IBA at 1000 ppm in August. Similarly, Abdullah *et al.* (2006) revealed that the guava species is amenable for clonal propagation by mature stem cutting and recorded maximum (32.7) number of primary root in guava cuttings treated with 0.8 per cent IBA solution which was followed by 0.4 per cent IBA treatment and lowest (58.3) was in cuttings without IBA treatment. Sarkar and Ghosh (2006) reported that air layers of guava prepared during June and July showed maximum number of primary and secondary roots in alluvial zone.

Kumar *et al.* (2007) reported maximum (14.9 cm) shoot length and number of leaves (12.7) in mid-June patch budded guava plants which was followed by end-June. Patel *et al.* (2007) reported maximum (36.89 cm) length of sprouts, number of leaves/plant (29.67) and leaf width (5.59 cm) with patch budding in cultivar

Allahabad Safeda among different cultivars of guava under mid hills of Meghalaya. Similarly, Babu *et al.* (2009) recorded highest (12.6 cm) sprout length at two months after patch budding. Results of Rymbai *et al.* (2012) revealed that maximum (10.80) number of primary and secondary (22.44) roots, length of longest (10.78 cm), fresh (2.72 g) and dry (0.51 g) root weight, establishment percentage (83.33 per cent), number of leaves (6.67) at 45 days after transplanting (DAT) and (13.83) at 60 DAT and minimum (8.67) number of days for buds sprouts were recorded in air layered plants under open and poly house nursery in guava during 2008-09.

2.3 Influence of Soil Media and Time on Propagation

Results of Copes (1977) revealed that soil media containing vermiculite + perlite (2:1), vermiculite + perlite (1:1) and vermiculite + perlite (1:2) resulted in rooting percentage 73 per cent, 50 per cent and 62 per cent respectively in Douglas- fir cuttings. Caldwell *et al.* (1988) reported that highest (88 per cent) percentage of rooting was obtained with cuttings from apical and middle (66 per cent) portions of current season's growth collected in mid-June to mid- July and also observed that cuttings rooted in vermiculite were given higher root quality ratings than those in perlite. In case of *Acacia nilotica* better germination and growth parameters were produced by nursery mixture of Soil: Sand: FYM in 2:1:1 (Bahuguna and Pyarelal, 1990). Pyarelal and Karnataka (1993) reported that the seedlings of *Quercus leucotrichophora* could be raised by sowing the seeds in soil mixture consisting soil,sand and farm yard manure in 2:1:1 ratio in nursery for better germination and growth parameters.

Sudhakar *et al.* (1995) reported that rooting medium containing sand, soil and farm yard manure in the ratio of 1:1:1 was significantly better than the rooting medium with sand and soil in the ratio of 1:1. The seedlings were superior to other media tested. The seedling vigour index leaf area, number of branches and shoot: root ratio were superior in the nursery mixture of soil: sand: FYM (1:1:1) compared to other mixtures.

Costa *et al.* (2008) reported that media containing vermiculite (soil + organic compost + vermiculite, 1:1:1 v/v) showed the best results with respect to leaf fresh mass (5.028 g), leaf dry mass (0.662 g), root fresh mass (2.562 g) and root dry mass (0.206 g) in papaya. Isfendiyaroglu *et al.* (2009) reported that perlite + vermiculite (1:1 v/v) resulted in rooting (95 per cent) in both years of experiment with highest (10.8) mean number of roots, root length (47 mm), root dry weight (382 mg), root fresh weight (48.3 mg) and number of secondary roots (13) in olive cuttings. Gautam *et al.* (2010) revealed significantly higher root induction (90±3.87 per cent) in vermiculite followed by sand (50±4.16 per cent). Root formation was retarded when soil was used as potting medium (8.02±1.85 per cent).

Results of Ali (2011) revealed that substrate containing vermiculite as media constituent resulted in maximum increase (25.40 cm) in plant height, leaf area index (2.95 cm^2), number of leaves (13.10) per plant and shoot dry weight (2.81 g) while, media containing perlite resulted in maximum increase (24.10 cm) in plant height, leaf area index (2.08 cm^2), number of leaves (12.60) per plant and shoot dry weight

(2.32 g) in Dahlia. Sharma *et al.*, 2012 reported that out of various potting mixtures, the potting mixture containing soil: sand and FYM (1:1:1 v/v/v) gave maximum height and survival of strawberry plantlets. 100 per cent survival was obtained in the potting mixture containing soil, sand and FYM (1:1:1 v/v/v) which was superior to all other treatments.

Chapter 3

Methodology

The present investigation entitled "Propagation studies in guava (*Psidium guajava* L.) cv. L-49 under Jammu subtropics" was carried out in the experimental farm of Division of Fruit Science, Faculty of Agriculture, SKUAST –J, Udheywalla, Jammu during 2012-13. The methodology adopted and techniques employed during the course of investigation are described in this chapter.

3.1 Climate and Weather Conditions of Experimental Site

Udheywalla is situated in the sub - tropical zone at latitude of 32.40 °N and longitude of 74.54°E. The altitude of the place is 300 meters from mean sea level. Annual precipitation is about 1200 mm. The mean annual maximum and minimum temperature was 29.60 °C and 16.70 °C, respectively. The winter months experience mild to severe cold and temperature ranges from 6.5°C to 21.7 °C. December is the coldest month and minimum temperature goes low as 4.0 °C, however, the maximum, minimum temperature and evaporation rate rises from March onwards.

Table 1: Meteorological Observation on Weekly Intervals during the Period of Investigation 2012-13

Met. Week (1)	*Date and Month* (2)	*Rainfall (mm)* (3)	*R H 1 (mor.)* (4)	*R H 1 (Afternoon)* (5)	*Max. Temp. (°C)* (6)	*Min. Temp. (°C)* (7)
22.	28-3 June	0.0	54	17	42.7	21.4
23.	4-10	7.0	61	25	38.7	22.3
24.	11-17	0.0	55	19	42.0	22.6
25.	18-24	2.2	58	27	42.1	24.7
26.	25-1 July	3.6	59	31	39.9	23.9
27.	2-8	130.4	74	55	36.6	22.5
28.	9-15	19.8	76	60	34.6	24.7

Contd...

Table 1–*Contd...*

Met. Week (1)	*Date and Month* (2)	*Rainfall (mm)* (3)	*R H 1 (mor.)* (4)	*R H 1 (Afternoon)* (5)	*Max. Temp. (°C)* (6)	*Min. Temp. (°C)* (7)
29.	16-22	6.2	71	57	36.4	25.6
30.	23-29	119.8	82	68	34.9	24.5
31.	30-5 Aug	165.6	90	73	32.0	24.4
32.	6-12 Aug	10.2	80	62	33.4	24.8
33.	13-19	64.2	85	67	34.0	25.5
34.	20-26	363.9	94	80	31.0	23.6
35.	27-2 Sept.	18.0	79	65	33.7	24.1
36.	3-9	38.0	82	71	32.1	22.8
37.	10-16	159.8	84	66	32.9	22.6
38.	17-23	25.7	78	62	31.0	21.4
39.	24-30	0.0	76	50	33.0	22.6
40.	1-7 Oct	0.0	81	43	33.0	23.0
41.	8-14	0.0	76	39	32.0	16.6
42.	15-21	0.2	81	42	28.9	14.1
43.	22-28	0.0	78	54	28.1	11.6
44.	29-4 Nov	0.0	77	41	28.9	11.5
45.	5-11	0.0	76	40	27.4	10.7
46.	12-18	0.0	80	41	25.9	9.9
47.	19-25	0.0	79	36	24.7	8.3
48.	26-2 Dec	5.4	82	47	21.9	6.7
49.	3-9	0.0	88	41	23.1	4.6
50.	10-16	34.8	88	55	19.7	7.8
51.	17- 23	0.0	90	53	20.1	5.1
52.	24-31	34.4	91	75	13.3	6.3
1.	1-7 Jan	0.0	93.6	79.0	10.0	5.2
2.	8-14	0.6	90.9	63.0	17.6	4.6
3.	15-21	40.8	90.0	63.3	17.5	5.6
4.	22-28	0.0	86.1	50.1	19.3	2.6
5.	29-4 Feb	19.8	87.9	59.1	20.2	7.6
6.	5-11	44.4	92.1	60.4	20.1	4.8
7.	12-18	10.8	90.7	63.4	19.9	7.5
8.	19-25	65.8	93.0	72.7	19.6	8.4
9.	26-4 Mar	35.8	85.9	57.7	23.3	8.3
10.	5-11	0.0	78.4	53.3	27.8	10.9
11.	12-18	5.7	88.6	55.4	25.5	11.9
12.	19-25	35.7	80.0	46.9	27.7	13.1
13.	26-1 Apr	0.0	89.4	49.4	28.4	12.6

3.2 Nutrient Status of the Soil Used in the Experiment

In order to study the physico-chemical characteristics of experimental soil used for propagation media in polybags, composite soil samples were collected from the soil used in the experiment with the help of khurpi from soil surface upto 15 cm depth one month prior to the experiment.

Table 2: Physico-chemical Analysis of Experimental Soil

Particulars	*Contents*	*Method Used*
	Mechanical analysis	
Sand (per cent)	60.0	
Silt (per cent)	19.0	International Dispersion Method
Clay (per cent)	21.0	(Piper, 1966)
	Chemical analysis	
pH	6.7	1:2 soil-water suspension with Beckman glass electrode pH meter (Jackson, 1967)
Electrical Conductivity (dS^{-1})	0.20	1:2 soil- water suspension with Systronic Conductivity meter (Jackson, 1973)
Organic Carbon (per cent)	0.51	Walkley and Black method (1934)
Available Nitrogen (kg ha^{-1})	225.5	Alkaline potassium permanganate method (Subbiah and Asija, 1956)
Available Phosphours (kg ha^{-1})	13.84	Olsen *et al.* (1954)
Available Potassium (kg ha^{-1})	138.0	Ammonium acetate method (Jackson, 1967)

The data presented in the Table 2 indicated that the texture of the soil was sandy loam and soil was neutral in reaction. The available nitrogen was in low range with available phosphorus and potassium in medium range.

3.3 Experimental Details

The experiments were conducted to standardize most suitable method for guava propagation under Jammu sub-tropics and to standardize soil media and time for guava propagation. In the first experiment four different propagation methods *viz.* , patch budding, wedge grafting, air layering and cutting were tried on 15th to 21st of four different months (June, July, August and September). Thirty seedlings raised from desi guava seeds of one year were selected for each propagation method on every date with 10 seedlings per replicate, therefore, 240 seedlings raised rootstock, 120 softwood cuttings and 120 shoots were air layered in the experiment I. In the second experiment, soft wood cuttings of guava were quickly dipped in growth regulator *i.e.* IBA 4000 ppm (Kareem *et al.*, 2013) of uniform dose on 15th to 21st of different months (June, July, August and September) and planted in polybags at depth of 6-8 cm (Ullah *et al.*, 2005) with different soil media *viz.* , vermiculite, perlite, sand, soil and FYM, to increase sprouting percentage in minimum possible time.

3.3.1 Experiment I

To standardize most suitable method for guava propagation under Jammu sub-tropics.

A. Propagation Methods

1. Cutting
2. Air-layering
3. Patch budding
4. Wedge grafting

B. Time of Propagation

1. 15th to 21st June
2. 15th to 21st July
3. 15th to 21st August
4. 15th to 21st September

3.3.1.1 Site of Experiment

Research orchard of Fruit Science, Faculty of Agriculture, Sher-e-Kashmir University of Agricultural Sciences and Technology of Jammu, Udheywalla.

Table 3: Planting Details

1.	Total number of treatments	16
2.	Number of replications	3
3.	Total number of plants per treatment	10
4.	Design of Experiment	Factorial RBD

3.3.1.2 Experimental Details

The methodology adopted during propagational studies is described under relevant heads.

3.3.1.2.1 Seedlings for Rootstock

The seedlings used for the propagational studies were provided by Division of Fruit Science, Faculty of Agriculture, Sher-e-Kashmir University of Agricultural Sciences and Technology of Jammu. These seedlings were raised from locally (Desi) available guava seeds and were grown in poly bags. The seedlings were about one year old and of pencil thickness at the time of propagation. Seedlings of healthy appearance were used for the propagational studies. They were used as rootstock after attaining a stem diameter of 0.5 to 1.0 cm.

3.3.1.2.2 Selection and Preparation of Budwood for the Scion Shoot

The budwood for the scion was collected from fifteen year old guava trees cultivar L-49 planted in the Experimental Research orchard of Fruit Science, Faculty of Agriculture, Sher-e-Kashmir University of Agricultural Sciences and Technology of Jammu, Udheywalla. The scion shoots were severed from the mother tree on the same date of budding/grafting operation. For budding, the leaves of the scion were removed before severing the bud sticks from the mother plants. The budwood was kept moist by wrapping in moist gunny bags to avoid desiccation of buds due to hot weather until the time of budding operation. Bud sticks with atleast 3-4 buds were used for grafting.

3.3.1.3 Procedures/Methodology Adopted for different Propagation Methods

Four propagation methods (*i.e.* cutting, air layering, patch budding and wedge grafting) were tried on four dates *i.e.* on 15th to 21st of every month (June, July, August and September) to find out most suitable method and time for propagation of guava under Jammu sub - tropics.

3.3.1.3.1 Method Adopted for the Preparation of Guava Softwood Cutting

Soft wood cuttings of guava cv. L-49 were made by removing immature branches from the selected plants which were of 4 to 6 months old. Leaves were removed from only that part which was to be buried in propagation media. Cuttings of uniform size of 15 cm long with diameter 0.4 to 0.7 cm having 4 buds and 2 cut leaves were made by giving basal cut at 0.3 cm below a bud and upper cut about 1.0 cm above the bud. The prepared cuttings were treated with 4000 ppm IBA (Indole Butyric Acid). Basal ends at least 2-3 cm were quickly dipped in required strength of hormone. After treatment with specified growth regulator, the treated cuttings were planted in well prepared black polybags at a depth of 6-8 cm containing soil, sand and FYM in ratio 1:1:1 (Shanker, 1999).

3.3.1.3.2 Method Adopted for Guava Air Layering

In air layering, the branches of one or two years of age and about a lead pencil thickness were selected. The selected plants were healthy, well matured, uniform and vigorous. In selected shoots, a ring of bark about 2-2.5 cm was removed by girdling carefully by giving two circular cuts about 45-60 cm below the top end of a shoot and the exposed portion was rubbed without causing any injury to the underlying wood. Handful of moss grass dipped in IBA was applied evenly around the wounded barkless area. A polyethylene of 150 gauges was then wrapped. These layer shoots were detached from the parent plants after 90 days of layering, some of the successful rooted layers were transplanted in a field and some were taken to laboratory for data observation (Shanker, 1999).

3.4.1.3.3 Method Adopted for Guava Patch Budding

In patch budding, rootstocks of one to one and half years old of about lead pencil thickness were selected. At about 10 -12 cm above the ground surface a horizontal cut of 1.25 to 1.80 cm length on the rootstock deep enough to cut the bark only was made. About 2.5 to 3.75 cm apart from the horizontal cut another cut of the same size as the previous one was made. Two horizontal cuts at 2.5 to 3.75 cm apart were made with the help of patch budding knife. These parallel cuts were joined by two

Plate 1: Air layering in guava (*Psidium guajava* L.) Plants.

Plate 2: Patch Budding in Guava (*Psidium guajava* L.) Plants.

vertical cuts connecting the ends of horizontal cuts.Thus a rectangular patch of bark was removed from the stock plant.

From the selected scion branch, a rectangular piece of bark containing one eye of same size as given on the stock with the consideration that the bud lies in the centre of the bark patch was removed. The scion patch of bark was placed on the stock from where a similar patch of bark was removed. The bud was tied with polythene film leaving room 0.62 cm for the bud to resume its growth. The operation was performed as quickly as possible to avoid air exposure of stock and scion components (Shanker, 1999).

3.3.1.3.4 Method Adopted for Guava Wedge Grafting

Seedlings raised for rootstock in the nursery for approximately 1 year, attaining a stem diameter of 0.5 to 1.0 cm were picked up for wedge grafting. The scion shoots 15 to 18 cm long of pencil thickness (0.5- 1.0cm) with 3 to 4 healthy buds were used for grafting. Selected scion shoots were defoliated on the mother plants. At the same time, the apical growing portion of selected shoots was also beheaded, which helps in forcing the dormant buds to swell. In this way, the buds on the scion were made ready to start sprouting at the time of grafting. After selection of the scion material, rootstock (seedling) was headed back, leaving 15 to 18 cm long stem above the polyethylene bag. The beheaded rootstock was split open about 4.0 to 4.5 cm deep through the centre from cut end of the rootstock with grafting knife. A wedge shaped cut, slanting from both the sides (4.0-4.5 cm long) was made on the lower portion of the scion shoot. The scion stick was inserted into the split of the stock and pressed properly so that cambium tissues of rootstock and scion should come in contact with each other. The stock and scion combination was then tied with the help of 150 gauge, 2 cm wide and 25 to 30 cm long polyethylene strip and covered with polytube for one week (Shanker, 1999).

3.3.2 Experiment II

To standardize the soil media and time of guava propagation through cuttings under Jammu sub tropics.

The different types of soil media used in this experiment for raising guava cuttings, softwood cuttings of cv. L-49 were used in this experiment. The cuttings were made of uniform size 15 cm long, having four buds and two cut leaves and were planted in different combinations of propagation media on four dates *i.e.* on 15th to 21st of four different months (June, July, August and September).

A. Soil Media

1. Vermiculite + FYM (1:1:1)
2. Perlite + FYM (1:1)
3. Sand + FYM (1:1)
4. Vermiculite + Sand + FYM (1:1:1)
5. Perlite + Sand + FYM (1:1:1)

B. Time of Propagation

1. 15th to 21st June
2. 15th to 21st July
3. 15th to 21st August
4. 15th to 21st September

3.3.2.1 Site of Experiment

Research orchard of Fruit Science, Faculty of Agriculture, Sher-e-Kashmir University of Agricultural Sciences and Technology of Jammu, Udheywalla.

Table 4: Planting Details

1.	Total number of treatments	20
2.	Number of replications	3
3.	Total number of plants per treatment	10
4.	Design of Experiment	Factorial RBD

3.3.2.2 Treatment Details

T1: Vermiculite + FYM 1:1

T2: Perlite + FYM 1:1

Plate 3: Wedge Grafting with Polycap Covering of (*Psidium guajava* L.) Seedlings

T3: Sand + FYM 1:1

T4: Vermiculite+Sand+FYM 1:1:1

T5: Perlite +Sand +FYM 1:1:1

3.3.2.3 Filling of Polybags

The polybags were filled with different combinations of propagation media. Container of 1kg was taken and equal amount of vermiculite and FYM was mixed thoroughly. In the same manner other proportions of the soil media were prepared and filled into polybags and cuttings treated with IBA 4000 ppm were planted into them at the depth of 6-8 cm.

3.3.2.4 Planting of Cuttings

Softwood cuttings of guava cv. L-49 were brought from experimental guava orchard of Fruit Science Division of Fruit Science, Faculty of Agriculture, SKUAST-J. The cuttings were made of uniform size 15 cm long with diameter 0.4 to 0.7 cm, having four buds and two cut leaves. Cuttings were treated with uniform dose of IBA 4000 ppm. The basal ends of the cuttings (2-3 cm) were immersed in the IBA solution for 5 min. The treated cuttings were planted in polyethylene bags at a depth of 6-8 cm. The polyethylene bags contain mixture of propagation media in different proportions.

3.4 Aftercare of Propagated Plants/Intercultural Operations

3.4.1 Shading

Shading is done to protect the plants in polybags from scorching sun and low temperature during winter

3.4.2 Staking

Staking is done not only to support the plant but also prevent the plants from blowing away by strong wind.

3.4.3 Watering and Feeding

Daily watering to seedling is needed for a week. It is better if liquid fertilizer or foliar feeding is followed for efficient absorption of nutrients and later better nutrition of plants.

3.5 Observations Recorded

The biometrical observations were recorded on five randomly selected plants of each replication to assess the vegetative characters *i.e.* per cent cutting/layering/ bud/graft take success, number of days to sprout, percentage sprouting, plant height, number of new shoots per plant, number of new leaves per plant, stem thickness of cutting/layered plants (cm), scion and stock diameter of budded/ grafted plants, average leaf area and for other observations *i.e.* average leaf fresh weight (g), average leaf dry weight (mg) and chlorophyll percentage plants were

Plate 4: Polybags Filled-in with Soil Media of different Composition.

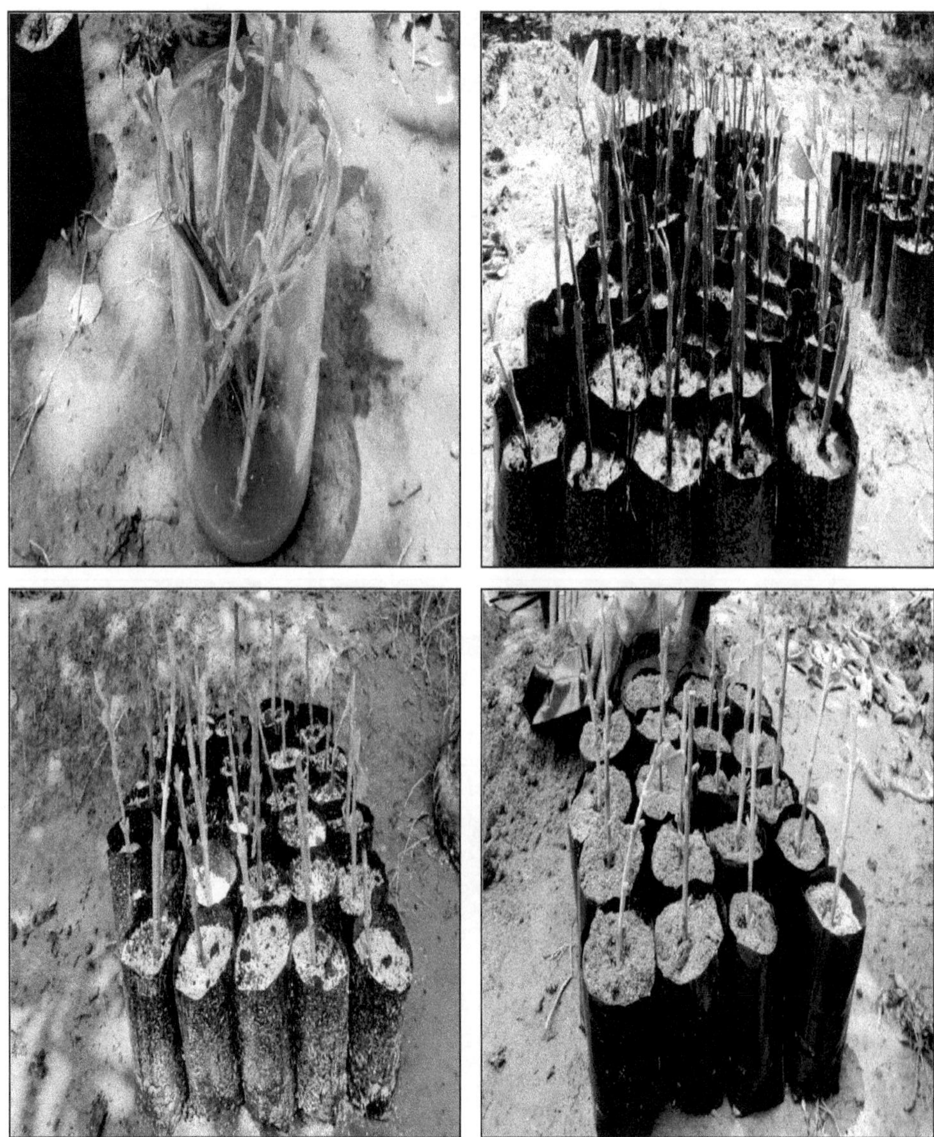

Plate 5: View of Guava (*Psidium guajava* L.) Softwood Cutting Treated with IBA and Planted in Polybags Containing Soil Media.

brought to the laboratory of Division of Fruit Science, FOA, Chatha, SKUAST-J for recording the data.

3.5.1 Observations Recorded in the Experiment I

3.5.1.1 Per cent Cutting/Layering/Bud/Graft take Success

The data on the per cent cuttings/layers/buds/graft take success were recorded after 90 days of planting the cuttings/layering/budding/grafting by taking the ratio of number of plants sprouted to the number of plants propagated and multiplied by 100.

$$\text{Per cent success} = \frac{\text{Number of plants survived}}{\text{Number of plants propagated}} \times 100$$

3.5.1.2 Percentage Sprouting

The data regarding the per cent sprouting were recorded after 90 days of planting cuttings/layering/budding/grafting by taking the ratio of number of plants sprouted to the number of plants planted and multiplied by 100.

$$\text{Per cent sprouting} = \frac{\text{Number of plants sprouted}}{\text{Number of plants propagated}} \times 100$$

3.5.2 Observation Recorded in II Experiment

The plants propagated by cuttings and planted in polybags containing soil media in different proportions were observed for the data on various parameters *viz.*, Observation on percentage of rooted cuttings, plant fresh weight (g), plant dry weight (g), number of primary roots per rooted cutting, number of secondary roots per rooted cutting, length of primary root, root fresh weight (g), root dry weight (mg), were taken after 90 days of planting cuttings.

3.5.2.1 Percentage of Rooted Cuttings

The data on the per cent rooted cuttings were recorded after 90 days of planting cuttings by taking the ratio of number of cuttings rooted to the number of cuttings planted and multiplied by 100.

$$\text{Per cent rooted cuttings} = \frac{\text{Number of cuttings rooted}}{\text{Number of cuttings planted}} \times 100$$

3.5.2.2 Plant Fresh Weight (g)

After washing the soil ball the data on average fresh weight of plant were recorded 90 days after planting the cutting in the propagation media. The weight was measured with the help of weighing balance and expressed as average fresh weight of plants in gram.

3.5.2.3 Plant Dry Weight (g)

After washing the soil ball the data on dry weight of plant were recorded 90 days after planting the cutting in propagation media. The dry weight was measured after oven drying of plant for 25 minute at 73ºC with the help of weighing balance and expressed as average dry weight of plant in gram.

3.5.2.4 Number of Primary Roots per Rooted Cutting

After washing the soil ball the data on number of primary roots per cutting were recorded 90 days of planting the cuttings in propagation media. Three guava plants propagated through softwood cuttings were randomly selected from each replication and expressed as number of primary roots per rooted cuttings.

3.5.2.5 Number of Secondary Roots per Rooted Cutting

After washing the soil ball the data on number of secondary roots per cutting were recorded after 90 days of planting cuttings in propagation media. Three guava plants propagated through softwood cuttings were randomly selected from each replication and expressed as number of secondary roots per rooted cuttings.

3.5.2.6 Length of Primary Root

After washing the soil ball the data on length of primary root were recorded after 90 days of planting cuttings in propagation media.Three guava plants propagated through softwood cuttings were randomly selected from each replication and root length was measured from the collar region to the tip of the longest root in centimeter.

3.5.2.7 Root Fresh Weight (g)

After washing the soil ball the data on fresh weight of roots were recorded 90 days after planting cutting in the propagation media. Three guava plants propagated through softwood cuttings were randomly selected from each replication and the root fresh weight was measured with the help of weighing balance and expressed as average fresh weight of roots in gram (gm).

3.5.2.8 Root Dry Weight (mg)

The data on dry weight of roots were recorded 90 days after planting cutting in propagation media. Three guava plants propagated through softwood cuttings were randomly selected from each replication and the root dry weight was measured after oven drying of plants for 25 minute at 73ºC and weighed with the help of weighing balance and expressed as average dry weight of roots in milligram.

3.5.3 Observations Recorded in both the Experiments I and II

3.5.3.1 Number of Days to Sprout

The data regarding the number of days taken to sprouting were calculated by observing the plants on alternate days from the day of planting cuttings/layering/budding/grafting and their mean was used to calculate the days taken for first sprout.

3.5.3.2 Plant Height (cm)

The data on average plant height of fully developed three representative plants was measured from the ground level to tip of the axis with the help of measuring scale at 30, 60, 90, 120, 150 and 180 days after planting cuttings/layering/budding/grafting and expressed as average plant height in centimeter (cm).

3.5.3.3 Number of New Shoots per Plant

The data on the average number of new shoots per plant of fully developed three representative plants were recorded at 30, 60, 90, 120, 150 and 180 days after planting cuttings/layering/budding/grafting.

3.5.3.4 Number of New Leaves per Plant

The data on the average number of new leaves per plant of fully developed three representative plants were recorded at 30, 60, 90, 120, 150 and 180 days after planting cuttings/layering/budding/grafting.

3.5.3.5 Stem Thickness of Cutting/Layered/Budded/Grafted Plants (cm)

The data on average radial growth (diameter) of stem of fully developed three representative cuttings and layered plants were recorded 4 cm above the ground level and budded and grafted plants were recorded 2 cm below the bud union or graft union at 30, 60, 90, 120, 150 and 180 days after planting cutting/layering. The diameter was measured with the help of digital vernier caliper and was expressed in centimeter (cm).

3.5.3.6 Average Leaf Area (cm^2)

The data on average leaf area was recorded 90 days after planting cutting/budding/layering/grafting. Average leaf area was calculated with the help of non destructive type of laser leaf area meter by taking five fully grown and matured leaves in each replication.

3.5.3.7 Average Leaf Fresh Weight (g)

After washing the soil ball the data on average fresh weight of leaves was recorded after 90 days of planting the cutting/budding/layering/grafting. The weight was measured from the weighing balance and expressed as average dry weight of roots in gram (gm).

3.5.3.8 Average Leaf Dry Weight (mg)

After washing the soil ball the data on average dry weight of leaves was recorded 90 days after planting the cutting/budding/layering/grafting. The dry weight was measured after drying in oven for 25 minute at 73°C from the weighing balance and expressed as average dry weight of leaves in milligram.

3.5.3.9 Chlorophyll Percentage

One gm of leaf sample was weighed and was ground in pestle mortar with 5 ml distilled water to paste. The contents were transferred to a centrifuge tube and the total volume was made upto10 ml with distilled water 0.5 ml from the tube was transferred to a tube containing 4.5ml of 80 per cent acetone. The contents were centrifuged at 4000 rpm for 15 min. The absorbance of the supernatant was measured at the wavelengths -645,663,490,638nm and the content of chlorophyll was calculated after 90 days.

3.6 Statistical Analysis

The data generated during the investigation were subjected to statistical analysis as prescribed by Panse and Sukhtame (2000).

Chapter 4

Experimental Results

The present investigation "Propagation studies in guava (*Psidium guajava* L.) cv. L-49 under Jammu sub-tropics were undertaken during 2012-13 at Fruit Plant Nursery, Faculty of Agriculture, Sher-e-Kashmir University of Agricultural Sciences and Technology of Jammu Udheywalla campus, Jammu. In the present studies two experiments were conducted *viz.*, studies to standardize most suitable method for guava propagation under Jammu sub-tropics and studies to find out best growing media and time for guava propagation through cuttings under Jammu sub-tropics. The observations for various parameters were recorded and data obtained were subjected to statistical analysis. The results of the investigation are presented under suitable headings as under.

4.1 Experiment I

To standardize most suitable method for guava propagation under Jammu sub-tropics.

4.1.1 Per cent Cutting/Layering/Bud/Graft take Auccess (after 90 days)

The data pertaining to the effect of time and methods of propagation on per cent cutting/layering/bud/graft take success is presented in Table 5. Perusal of data revealed that the time and methods of propagation showed significant effect on per cent cutting/layering/bud/graft take success of guava plants.

4.1.1.1 Influence of Time and Methods of Propagation on Per Cent Success in Guava cv. L-49

Data presented in Table 5 on per cent success among different time and methods of propagation recorded highest (67.74 per cent) average per cent success in patch budding which was followed by 63.19 in wedge grafting while, lowest (45.06 per cent) in air layering. At 90 DAP, highest (71.48 per cent) average per cent success was recorded during 15th-21st of August which was followed by (68.69 per cent) during 15th-21st September while, lowest (31.65 per cent) during 15th-21st of June.

Table 5: Influence of Time, Method of Propagation and their Interaction Effect on per cent Success of Guava cv. L-49 after 90 Days

Method of Propagation	*Time of Propagation*				
	June 15th-21st	*July 15th-21st*	*Aug 15th-21st*	*Sep 15th-21st*	*Mean*
Cutting	35.81 (34.30)	51.87 (61.80)	54.21 (65.80)	50.31 (59.23)	**48.05 (55.28)**
Air Layering	34.15 (31.69)	49.74 (58.17)	44.42 (49.02)	40.00 (41.34)	**42.08 (45.06)**
Patch budding	34.35 (31.88)	50.43 (59.17)	73.70 (92.07)	69.37 (87.61)	**56.96 (67.74)**
Wedge grafting	32.28 (28.74)	49.83 (58.36)	63.03 (79.04)	68.47 (86.57)	**53.40 (63.19)**
Mean	**34.15 (31.65)**	**50.46 (59.44)**	**58.84 (71.48)**	**50.46 (68.69)**	
Factors		*S.E. m (±)*		*C.D (0.05)*	
Method of propagation		0.99		2.87	
Time of propagation		0.99		2.87	
Method x Time		**1.98**		**5.73**	

Note: Figures in parenthesis indicate observed values and others are transformed values.

4.1.1.2 Interaction Effect of Time and Methods of Propagation on per cent Success in Guava cv. L-49

The data presented in Table 5 on interaction effect of time and methods of propagation recorded highest (92.07 per cent) per cent success in patch budding

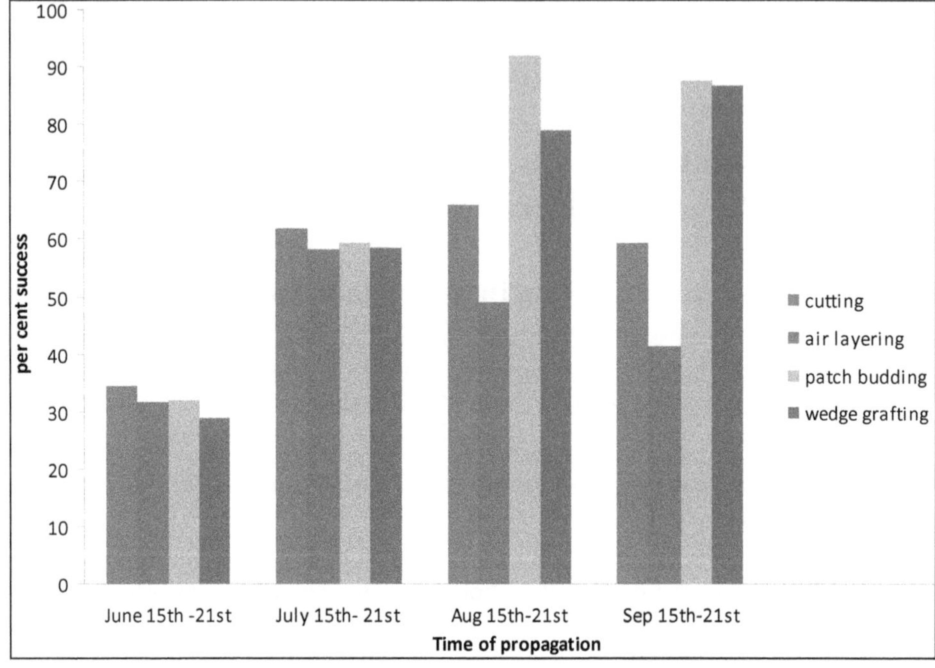

Figure 1: Influence of Time and Methods of Propagation on per cent Success in Guava cv. L-49 after 90 Days.

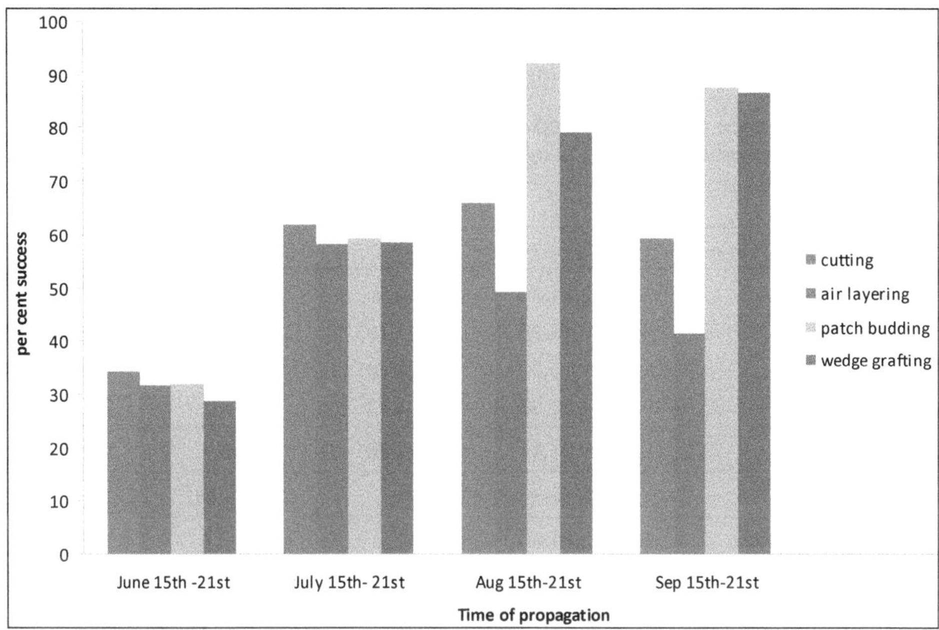

Figure 2: Influence of Time and Method of Propagation on per cent Success of Guava cv. L-49 after 90 Days.

during 15th-21st of August which was followed by (79.04 per cent) wedge grafting while, the lowest (49.02 per cent) in air layering. Similarly, during 15th-21st September highest (87.61 per cent) success was recorded in patch budding which was at par with (86.57 per cent) wedge grafting while, the lowest (41.34 per cent) in air layering.

Data pertaining to interaction effect of time and methods of propagation during 15th-21st of July recorded highest (61.80 per cent) per cent success in cuttings which was at par with (59.17 per cent) patch budding, (58.36 per cent) wedge grafting and (58.17 per cent) air layering. Similarly, during 15th-21st of June highest (34.30 per cent) success was recorded in cuttings which were at par with patch budding (31.88 per cent), air layering (31.69 per cent) and wedge grafting (28.74 per cent).

4.1.2 Number of Days to Sprout in Guava cv. L-49

The data related to effect of time and methods of propagation on number of days taken to sprout is presented in Table 6. A perusal of data revealed that both time and methods of propagation showed significant effect on number of days taken to sprouting of guava.

4.1.2.1 Influence of Time and Methods of Propagation on Number of Days to Sprout in Guava cv. L-49

Data presented in Table 6 on per cent success among different time and methods of propagation recorded minimum (14.65) days to sprout in air layering which was followed by (19.26) cuttings while, maximum (25.26) in patch budding. The effect of time of propagation was also significant with respect to days taken to sprout in guava, the minimum days taken to sprout (17.76) was observed during 15th-21st of August which was followed by (19.54) during 15th-21st of September while, the maximum (23.53) during 15th-21st of June.

Table 6: Influence of Time, Method of Propagation and their Interaction Effect on Number of Days taken to Sprouting of Guava cv. L-49

Method of Propagation	Time of Propagation				
	June 15th-21st	July 15th-21st	Aug 15th-21st	Sep 15th-21st	Mean
Cutting	19.74	21.71	16.90	18.70	**19.26**
Air Layering	18.70	15.38	11.75	12.76	**14.65**
Patch budding	28.00	25.00	23.68	24.35	**25.26**
Wedge grafting	27.69	23.68	18.72	22.34	**23.11**
Mean	**23.53**	**21.44**	**17.76**	**19.54**	

Factors	S.E. m (±)	C.D (0.05)
Method of propagation	0.47	1.37
Time of propagation	0.47	1.37
Method x Time	**0.95**	**2.75**

4.1.2.2 Interaction Effect of Time and Methods of Propagation on Number of Days taken to Sprout in Huava cv. L-49

The data presented in Table 6 on interaction effect of time and methods of propagation recorded minimum (11.75) days to sprout in air layering during 15th-21st of August which was followed by (16.90) cuttings while, maximum (23.68) days taken to sprout was recorded in patch budding. Similarly, during 15th-21st of September minimum (12.76) days taken to sprout was recorded in air layering which was followed by (18.70) in cuttings while, maximum (24.35) in patch budding.

Data pertaining to interaction effect of time and methods of propagation recorded minimum (15.38) days to sprout in air layering during 15th-21st of July which was followed by (21.71) cuttings while, maximum (25.00) in patch budding. Similarly, during 15th-21st of June minimum (18.70) days to sprout was observed in air layering which was followed by (19.74) in cuttings while, maximum (28.00) in patch budding.

4.1.3 Percentage Sprouting (after 90 days of planting/layering/budding/grafting)

The result obtained with respect to effect of various time and methods of propagation on percentage sprouting after 90 days is presented in Table 7. It is clear from the data that both time and methods of budding had significant effect on percentage sprouting of guava cv.L-49.

Table 7: Influence of Time, Method of Propagation and their Interaction Effect per cent Sprouting of Guava cv. L-49 after 90 Days

Method of Propagation	*Time of Propagation*				
	June 15th-21st	*July 15th-21st*	*Aug 15th-21st*	*Sep 15th-21st*	*Mean*
Cutting	43.06 (46.68)	45.69 (51.23)	65.25 (82.50)	62.76 (79.08)	**54.19 (64.87)**
Air Layering	41.59 (44.00)	52.41 (62.71)	63.76 (80.00)	55.45 (67.70)	**53.28 (63.60)**
Patch budding	57.46 (71.11)	64.79 (81.90)	82.02 (98.00)	75.56 (93.68)	**69.96 (86.17)**
Wedge grafting	57.44 (71.07)	64.52 (81.53)	69.20 (87.37)	73.44 (91.90)	**66.15 (82.97)**
Mean	49.86 (58.21)	56.85 (69.39)	70.05 (86.97)	66.80 (83.90)	

Factors	*S.E. m (±)*	*C.D (0.05)*
Method of propagation	1.12	3.23
Time of propagation	1.12	3.23
Method x Time	**2.23**	**6.45**

Note: Figures in parenthesis indicate observed values and others are transformed values.

4.1.3.1 Influence of Time and Methods of Propagation on Percentage Sprouting in Guava cv. L-49 after 90 Days

Data presented in Table 7 on per cent sprouting among different time and methods of propagation recorded maximum (86.17 per cent) average per cent sprouting in patch budding which was followed by (82.97 per cent) wedge grafting while, it was found minimum (63.60 per cent) in air layering. The effect of time of propagation was also significant with respect to per cent sprouting, the maximum (86.97 per cent) sprouting was observed during 15th-21st of July which was followed by (83.90 per cent) during 15th-21st of September while, minimum (58.21 per cent) during 15th-21st of June.

4.1.3.2 Interaction Effect of Time and Methods of Propagation on Percentage Sprouting in Guava cv. L-49 after 90 Days

The data presented in Table 7 on interaction effect of time and methods of propagation recorded highest (98.00 per cent) per cent sprouting in patch budding during 15th-21st of August which was followed by wedge grafting (87.37 per cent) while, lowest (80.00 per cent) in air layering. Similarly, during 15th-21st of September highest (93.68 per cent) per cent sprouting was observed in patch budding which was followed by wedge grafting (91.90 per cent) while, it was observed minimum (67.70 per cent) in air layering.

Data pertaining to interaction effect of time and methods of propagation recorded highest sprouting (81.90 per cent) in patch budding during 15th-21st of July which was at par with wedge grafting (81.53 per cent) while, lowest (51.23 per cent) in cuttings. Similarly, during 15th-21st June highest (71.11 per cent) per cent sprouting in patch budding was observed which was at par with wedge grafting (71.07 per cent) while, air layering recorded the lowest per cent sprouting (44.00 per cent).

4.1.4 Plant Height

The data pertaining to the effect of time and methods of propagation on plant height at 30, 60, 90, 120, 150 and 180 days after propagation in guava plants are presented in Table 8.

4.1.4.1 Plant Height at 30 DAP (cm)

The data obtained after 30 days on the influence of time and methods of propagation on plant height of guava cv. L-49 showed non significant differences presented in Table 8.

4.1.4.2 Plant Height at 60 DAP (cm)

The data presented in Table 8 revealed that both time and methods of propagation had significant effect on plant height in guava.

4.1.4.2.1 Influence of Time and Methods of Propagation on Plant Height (cm) in Guava cv. L-49

Data presented in Table 8 on average plant height as influenced by time and methods of propagation revealed that air layering recorded maximum (49.19 cm) average plant height which was followed by wedge grafting (21.02 cm) while, minimum (18.20 cm) in cuttings. After 60 DAP the maximum (29.61 cm) average plant height was observed during 15th-21st of August which was followed by (28.12 cm) during 15th-21st of September while, minimum (23.64 cm) during 15th-21st of June.

4.1.4.2.1 Interaction effect of time and methods of propagation on plant height (cm) of guava cv. L-49.

The data presented in Table 8 on interaction effect of time and methods of propagation showed maximum (53.60 cm) plant height in air layering during 15th-21st of August which was followed by wedge grafting (23.54 cm) while, the minimum (20.33 cm) in cuttings. Similarly, during 15th-21st of September maximum (48.53 cm) plant height was recorded in air layering which was followed by (23.53 cm) wedge grafting while, minimum (19.43 cm) in cuttings.

Data pertaining to interaction effect of time and methods of propagation during 15th-21st of July recorded maximum (48.17 cm) plant height in air layering which was followed by wedge grafting (18.86 cm) while, minimum (17.43 cm) in cuttings. Similarly, during 15th-21st of June maximum (46.47 cm) plant height was recorded in air layering which was followed by wedge grafting (16.90 cm) and the minimum (15.63 cm) in cuttings.

4.1.4.3 Plant Height at 90 DAP (cm)

It is apparent from the results presented in Table 8 that significant difference in both factors *i.e.* time and methods of propagation was observed with respect to plant height at 90 DAP.

Table 8: Influence of Time, Method of Propagation and their Interaction Effect on Plant Height of Guava cv. L-49

Time of Propagation	*Method of Propagation*														
	June 15th-21st	*July 15th-21st*	*Aug 15th-21st*	*Sep 15th-21st*	*Mean*	*June 15th-21st*	*July 15th-21st*	*Aug 15th-21st*	*Sep 15th-21st*	*Mean*	*June 15th-21st*	*July 15th-21st*	*Aug 15th-21st*	*Sep 15th-21st*	*Mean*
	After 30 Days					*After 60 Days*					*After 90 Days*				
Cutting	14.38	15.67	18.43	17.17	16.88	15.63	17.43	20.33	19.43	18.20	16.50	19.50	23.27	21.43	**20.12**
Air Layering	46.93	44.40	51.23	51.00	48.14	46.47	48.17	53.60	48.53	49.19	47.93	50.67	54.80	50.03	**50.86**
Patch budding	15.00	15.33	14.47	16.68	15.42	15.53	17.97	20.97	20.97	18.86	17.27	21.10	26.07	24.10	**22.13**
Wedge grafting	16.34	16.68	18.73	17.69	17.37	16.90	18.86	23.54	23.53	21.02	18.27	23.37	28.33	26.57	**24.13**
Mean	23.16	23.25	26.25	25.63		23.64	25.60	29.61	28.12		24.99	28.61	33.12	30.61	
Factors		*S.E. m (±)*		*C.D (0.05)*			*S.E. m (±)*		*C.D (0.05)*			*S.E. m (±)*		*C.D (0.05)*	
Method of propagation		N.S					0.16		0.48			0.24		0.71	
Time of propagation		N.S					0.16		0.48			0.24		0.71	
Method x Time		**N.S**					**0.33**		**0.96**			**0.50**		**1.42**	

Contd...

Table 8–*Contd...*

Time of Propagation	*Method of Propagation*														
	June 15th-21st	*July 15th-21st*	*Aug 15th-21st*	*Sep 15th-21st*	*Mean*	*June 15th-21st*	*July 15th-21st*	*Aug 15th-21st*	*Sep 15th-21st*	*Mean*	*June 15th-21st*	*July 15th-21st*	*Aug 15th-21st*	*Sep 15th-21st*	*Mean*
	After 120 Days					*After 150 Days*					*After 180 days*				
Cutting	18.47	20.40	23.16	21.50	20.88	19.00	21.27	24.08	22.07	21.60	22.37	25.70	28.69	27.39	**26.04**
Air Layering	50.90	52.30	57.83	52.10	53.28	51.90	54.23	60.00	54.97	52.27	52.70	55.10	61.89	58.47	**56.79**
Patch budding	19.25	21.87	27.40	26.40	23.73	21.32	24.90	31.07	29.87	26.79	23.52	30.86	33.52	30.76	**29.66**
Wedge grafting	21.71	23.73	28.90	28.90	25.48	23.72	27.27	32.77	29.57	28.33	26.91	30.01	35.40	32.35	**31.17**
Mean	27.58	29.57	33.63	32.31		28.98	31.92	36.98	34.12		31.40	35.17	39.87	37.24	
Factors		*S.E. m (±)*		*C.D (0.05)*			*S.E. m (±)*		*C.D (0.05)*			*S.E. m (±)*		*C.D (0.05)*	
Method of propagation		0.32		0.93			0.22		0.62			0.34		0.97	
Time of propagation		0.32		0.93			0.22		0.62			0.34		0.97	
Method x Time		**0.64**		**1.86**			**0.43**		**1.25**			**0.67**		**1.94**	

4.1.4.3.1 *Influence of Time and Methods of Propagation on Plant Height (cm) in Guava cv. L-49*

Data presented in Table 8 on average plant height at 90 DAP as influenced by time and methods of propagation recorded maximum (50.86 cm) average plant height in air layering which was followed by wedge grafting (24.13 cm) while, minimum (20.12 cm) in cuttings. Similarly, during 15th-21st August, the plant height was recorded maximum (33.12 cm) which was followed by (30.61 cm) during 15th-21st of September while, it was found minimum (24.99 cm) during 15th-21st June.

4.1.4.3.1 *Interaction Effect of Time and Methods of Propagation on Plant Height (cm) in Guava cv. L-49*

The data presented in Table 8 on interaction effect of methods and time of propagation recorded maximum (54.80 cm) plant height in air layering during 15th-21st of August which was followed by wedge grafting (28.33 cm) while, minimum (23.27 cm) in cuttings. Similarly, during 15th-21st of September maximum (50.03 cm) plant height was recorded in air layering which was followed by (26.57 cm) wedge grafting while, minimum (21.43 cm) in cuttings.

Data pertaining to interaction effect of time and methods of propagation during 15th-21st July recorded maximum (50.67 cm) plant height in air layering which was followed by wedge grafting (23.37 cm) while, minimum (19.50 cm) in cuttings. Similarly, data obtained during 15th-21st of June recorded maximum (47.93 cm) plant height in air layering which was followed by wedge grafting (18.27 cm) while, minimum (16.50 cm) in cuttings.

4.1.4.4 Plant Height at 120 DAP (cm)

The data pertaining to plant height at 120 DAP of guava cv. L-49 as influenced by time and methods of propagation are presented in Table 8. It is clear from the data that results showed significant effect on the plant height of guava at 120 DAP.

4.1.4.4.1 *Influence of Time and Methods of Propagation on Plant Height (cm) in Guava cv. L-49*

The data presented in Table 8 on average plant height at 120 DAP as influenced by time and methods of propagation showed maximum (53.28 cm) average plant height in air layering which was followed by (25.48 cm) wedge grafting while, minimum (20.88 cm) in cuttings. Similarly, the maximum (33.63 cm) average plant height was observed when propagation was performed during 15th-21st of August which was followed by (32.31 cm) during 15th-21st of September while, minimum (27.58 cm) was recorded during 15th-21st of June.

4.1.4.4.2 *Interaction Effect of Time and Methods of Propagation on Plant Height (cm) in Guava cv. L-49*

The data presented in Table 8 on interaction effect of time and methods of propagation recorded maximum (57.83 cm) plant height in air layering during 15th-21st of August which was followed by wedge grafting (28.90 cm) while it was found minimum (23.16 cm) in cuttings. Similarly, during 15th-21st of September maximum (52.10 cm) plant height was recorded in air layering which was followed by wedge grafting (28.90 cm) while, minimum (21.50 cm) in cuttings.

Data pertaining to interaction effect of time and methods of propagation during 15th-21st of July showed maximum (52.30 cm) plant height in air layering which was followed by (23.73 cm) while, it was found minimum (20.40 cm) in cuttings. Similarly, data recorded during 15th-21st of June showed maximum (50.90 cm) plant height in air layering which was followed by wedge grafting (21.71 cm) while, minimum (18.47 cm) in cuttings.

4.1.4.5 Plant Height at 150 DAP (cm)

The data pertaining to plant height at 150 DAP of guava cv. L-49 as influenced by time and methods of propagation are presented in Table 8. It is clear from the data that results showed significant effects on the plant height of guava at 150 DAP.

4.1.4.5.1 Influence of Time and Methods of Propagation on Plant Height (cm) in G

Data presented in Table 8 on average plant height at 150 DAP as influenced by time and methods of propagation showed the maximum (55.27 cm) average plant height in air layering which was followed by (28.33 cm) wedge grafting while, minimum (21.60 cm) in cuttings. Similarly, maximum (36.98 cm) average plant height was observed during 15th-21st of August which was followed by (34.12 cm) during 15th-21st September propagated plants while, minimum (28.98 cm) during 15th-21st of June

4.1.4.5.2 Interaction Effect of Time and Methods of Propagation on Plant Height (cm) in Guava cv. L-49

The data presented in Table 8 on interaction effect of time and methods of propagation recorded maximum (60.00 cm) plant height in air layering during 15th-21st of August which was followed by (32.77 cm) wedge grafting while it was found minimum (24.08 cm) in cuttings. Similarly data obtained from different propagation methods during 15th-21st of September showed maximum (54.97 cm) plant height in air layering which was followed by wedge grafting (29.57 cm) while, minimum (22.07 cm) in cuttings.

Data pertaining to interaction effect of time and methods of propagation during 15th-21st of July showed maximum (54.23 cm) plant height in air layering which was followed by wedge grafting (27.27 cm) while, it was found minimum (21.27 cm) in cuttings. Similarly, data recorded during 15th-21st of June showed maximum (51.90 cm) plant height in air layering which was followed by wedge grafting (23.72 cm) while, minimum (19.00 cm) in cuttings.

4.1.4.6 Plant Height at 180 DAP (cm)

The data pertaining to plant height at 180 DAP of guava cv. L-49 as influenced by time and methods of propagation are presented in Table 8. It is clear from the data that results showed significant effect on the plant height of guava at 180 DAP.

4.1.4.6.1 Influence of Time and Methods of Propagation on Plant Height (cm) in Guava cv. L-49

The data presented in Table 8 on average plant height at 180 DAP as influenced by time and methods of propagation showed the maximum (56.79 cm) average plant height in air layering which was followed by wedge grafting (31.17 cm) while, minimum (26.04 cm) in cuttings. Similarly, the maximum (39.87 cm) average plant height was observed during 15th-21st of August which was followed by (37.24 cm) during 15th-21st September propagated plants while, minimum average (31.40 cm) during 15th-21st of June.

4.1.4.6.2 Interaction Effect of Time and Methods of Propagation on Plant Height (cm) in Guava cv. L-49

The data presented in Table 8 on interaction effect of time and methods of propagation recorded maximum (61.89 cm) plant height in air layering during 15th-21st of August which was followed by wedge grafting (35.40 cm) while it was found minimum (28.69 cm) in cuttings. Similarly, during 15th-21st of September maximum (58.47 cm) plant height was recorded in air layering which was followed by wedge grafting (32.35 cm) while, minimum (27.39 cm) in cuttings.

Data pertaining to interaction effect of time and methods of propagation tried during 15th-21st of July showed maximum (55.10 cm) plant height in air layering which was followed by wedge grafting (30.01 cm) while, it was found minimum (25.70 cm) in cuttings. Similarly, during 15th-21st of June maximum (52.70 cm) plant height was recorded in air layering which was followed by wedge grafting (26.91 cm) while, minimum (22.37 cm) in cuttings.

4.1.5 Number of Shoots per Plant

The data pertaining to the effect of time and methods of propagation on number of shoots per plant at 30, 60, 90, 120, 150 and 180 days after propagation are presented in Table 9.

4.1.5.1 Number of Shoots per Plant at 30 DAP

The data obtained after 30 days on the influence of time and method of propagation on number of shoots of guava cv. L-49 showed non significant differences presented in Table 9.

4.1.5.2 Number of Shoots per Plant at 60 DAP

The data presented in Table 9 revealed that both time and methods of propagation had significant effect on number of shoots of guava.

4.1.5.2.1 Influence of Time and Methods of Propagation on Number of Shoots per Plant in Guava cv. L-49

The data presented in Table 9 on average number of shoots as influenced by time and methods of propagation recorded maximum (3.30) average number of shoots in wedge grafting which was followed by (3.19) patch budding while, it was found minimum (2.25) in cuttings. After 60 DAP the maximum (3.34) average

Table 9: Influence of Time, Method of Propagation and their Interaction Effect on Number of Shoots per Plant of Guava cv. L-49

Time of Propagation	*Method of Propagation*														
	June 15th-21st	*July 15th-21st*	*Aug 15th-21st*	*Sep 15th-21st*	*Mean*	*June 15th-21st*	*July 15th-21st*	*Aug 15th-21st*	*Sep 15th-21st*	*Mean*	*June 15th-21st*	*July 15th-21st*	*Aug 15th-21st*	*Sep 15th-21st*	*Mean*
	After 30 Days					*After 60 Days*					*After 90 Days*				
Cutting	1.01	1.48	1.81	1.25	1.39	1.98	1.97	3.04	2.01	2.25	2.95	3.82	3.97	3.89	**3.66**
Air Layering	1.70	2.01	1.77	2.08	1.88	2.00	2.03	3.06	2.38	2.37	2.92	3.38	3.47	3.38	**3.28**
Patch budding	1.80	1.10	1.33	1.03	1.07	2.38	3.37	3.56	3.45	3.19	2.98	3.86	4.05	3.96	**3.71**
Wedge grafting	1.18	1.03	1.50	1.80	1.38	2.44	3.49	3.69	3.57	3.30	3.17	3.98	4.07	3.98	**3.80**
Mean	1.17	1.40	1.60	1.53		2.20	2.71	3.34	2.85		3.00	3.71	3.88	3.80	

Factors	*S.E. m (±)*	*C.D (0.05)*	*S.E. m (±)*	*C.D (0.05)*	*S.E. m (±)*	*C.D (0.05)*
Method of propagation	N.S		0.08	0.22	0.01	0.01
Time of propagation	N.S		0.08	0.22	0.01	0.01
Method x Time	**N.S**		**0.15**	**0.44**	**0.01**	**0.03**

Contd...

Table 9–*Contd...*

Time of Propagation	*Method of Propagation*														
	June 15th-21st	*July 15th-21st*	*Aug 15th-21st*	*Sep 15th-21st*	*Mean*	*June 15th-21st*	*July 15th-21st*	*Aug 15th-21st*	*Sep 15th-21st*	*Mean*	*June 15th-21st*	*July 15th-21st*	*Aug 15th-21st*	*Sep 15th-21st*	*Mean*
	After 120 Days					*After 150 Days*					*After 180 days*				
Cutting	3.18	4.02	4.17	4.07	3.86	3.99	4.81	4.88	4.70	4.59	4.37	5.19	5.36	5.28	5.05
Air Layering	3.16	3.97	4.10	4.05	3.82	3.78	4.75	4.88	4.82	4.56	4.27	5.10	4.94	5.21	**4.88**
Patch budding	3.28	4.76	4.97	4.87	4.47	5.39	4.93	5.07	5.00	5.10	6.28	6.71	6.88	6.76	**6.66**
Wedge grafting	3.38	4.78	4.98	4.88	4.50	5.40	4.94	5.08	5.01	5.11	6.30	6.81	8.97	6.85	**7.23**
Mean	**3.25**	**4.38**	**4.57**	**4.47**		**4.64**	**4.86**	**4.98**	**4.88**		**5.30**	**5.95**	**6.54**	**6.02**	

Factors	*S.E. m (±)*	*C.D (0.05)*	*S.E. m (±)*	*C.D (0.05)*	*S.E. m (±)*	*C.D (0.05)*
Method of propagation	0.01	0.01	0.04	0.11	0.04	0.12
Time of propagation	0.01	0.01	0.04	0.11	0.04	0.12
Method x Time	**0.01**	**0.03**	**0.07**	**0.21**	**0.08**	**0.24**

number of shoots was observed during 15th-21st of August which was followed by (2.85) during 15th-21st September while, minimum (2.20) during 15th-21st of June.

4.1.5.2.2 Interaction Effect of Time and Methods of Propagation on Number of Shoots per Plant in Guava cv. L-49

The data presented in Table 9 on interaction effect of time and methods of propagation showed maximum (3.69) number of shoots in wedge grafting during 15th-21st of August which was at par with (3.56) patch budding while, minimum (3.04) in cuttings. Similarly, during 15th-21st of September maximum (3.57) number of shoots was recorded in wedge grafting which was at par with (3.45) patch budding while, minimum (2.01) in cuttings.

Data pertaining to interaction effect of time and methods of propagation during 15th-21st of July recorded maximum (3.49) number of shoots in wedge grafting which was at par with (3.37) patch budding while, minimum (1.97) in cuttings. Similarly, during 15th-21st of June maximum (2.44) number of shoots was observed in wedge grafting which was at par with (2.38) patch budding and (2.00) in air layering while, minimum (1.98) in cuttings.

4.1.5.3 Number of Shoots per Plant at 90 DAP

It is apparent from the results presented in Table 9 that significant difference in both factors *i.e.* time and methods of propagation was observed with respect number of shoots per plant at 90 DAP.

4.1.5.3.1 Influence of Time and Methods of Propagation on Number of Shoots per Plant in Guava cv. L-49

The data presented in Table 9 on average number of shoots at 90 DAP as influenced by time and methods of propagation recorded maximum (3.80) average number of shoots in wedge grafting, followed by patch budding (3.71) while, minimum (3.28) in air layering. Similarly, during 15th-21st of August, maximum (3.88) average number of shoots were recorded which was followed by (3.80) during 15th-21st of September while, minimum (3.00) during 15th-21st of June.

4.1.5.3.2 Interaction Effect of Time and Methods of Propagation on Number of Shoots per Plant in Guava cv. L-49

The data presented in Table 9 on interaction effect of methods and time of propagation showed that maximum (4.07) number of shoots was recorded in wedge grafting during 15th-21st of August which was at par with (4.05) patch budding while, minimum (3.47) in air layering. Similarly, during 15th-21st of September maximum (3.98) number of shoots was recorded in wedge grafting which was at par with (3.96) patch budding while, minimum (3.38) in air layering.

Data pertaining to interaction effect of methods and time of propagation during 15th-21st of July recorded maximum (3.98) number of shoots in wedge grafting which was at par with (3.86) patch budding while, minimum (3.38) in air layering. Similarly, during 15th-21st of June maximum (3.17) number of shoots was recorded in wedge grafting which was followed by (2.98) patch budding while, minimum (2.92) in air layering.

4.1.5.4 Number of Shoots per Plant at 120 DAP

The data pertaining to plant height at 120 DAP of guava cv. L-49 as influenced by time and methods of propagation are presented in Table 9. It is clear from the data that results showed significant effect on the number of shoots per plant in guava at 120 DAP.

4.1.5.4.1 Influence of Time and Methods of Propagation on Number of Shoots per Plant in Guava cv. L-49

The data presented in Table 9 on average number of shoots at 120 DAP as influenced by time and methods of propagation showed maximum (4.50) average number of shoots in wedge grafting which was followed by (4.47) patch budding while, it was found minimum (3.82) in air layering. Similarly, the maximum (4.57) average number of shoots was observed during 15th-21st of August which was followed by (4.47) during 15th-21st of September while, minimum (3.25) during 15th-21st of June.

4.1.5.3.2 Interaction Effect of Time and Methods of Propagation on Number of Shoots per Plant in Guava cv. L-49

The data presented in Table 9 on interaction effect of time and methods of propagation recorded maximum (4.98) number of shoots in wedge grafting during 15th-21st of August which was at par with (4.97) patch budding while, minimum (4.10) in air layering. Similarly, during 15th-21st of September maximum (4.88) number of shoots was recorded in wedge grafting which was at par with (4.87) patch budding while, minimum (4.05) in air layering.

Data pertaining to interaction effect of time and methods of propagation recorded maximum (4.78) number of shoots in wedge grafting during 15th-21st of July which was at par with (4.76) in patch budding while, minimum (3.97) in air layering. Similarly, during 15th-21st of June maximum (3.38) number of shoots was recorded in wedge grafting which was followed by patch budding (3.28) while, minimum (3.16) in air layering.

4.1.5.5 Number of Shoots per Plant at 150 DAP

The data pertaining to plant height at 150 DAP of guava cv. L-49 as influenced by methods of propagation are presented in Table 9. It is clear from the data that results showed significant effect on the number of shoots per plant in guava at 150 DAP.

4.1.5.5.1 Influence of Time and Methods of Propagation on Number of Shoots per Plant in Guava cv. L-49

The data presented in Table 9 on average number of shoots at 150 DAP as influenced by time and methods of propagation showed maximum (5.11) average number of shoots in wedge grafting which was at par with (5.10) patch budding while, minimum (4.56) in air layering. Similarly, the maximum (4.98) average number of shoots was observed during 15th-21st of August which was followed by (4.88) during 15th-21st of September while, it was recorded minimum (4.64) during 15th-21st of June.

4.1.5.5.2 Interaction Effect of Time and Methods of Propagation on Number of Shoots per Plant in Guava cv. L-49

The data presented in Table 9 on interaction effect of methods and time of propagation recorded maximum (5.08) number of shoots in wedge grafting during 15th-21st of August which was at par with (5.07) patch budding and (4.88) cuttings while, minimum (4.88) in cuttings. Similarly, the maximum (5.01) number of shoots was recorded in wedge grafting which was at par with (5.00) patch budding while, minimum (4.82) in air layering.

Data pertaining to interaction effect of time and methods of propagation showed maximum (4.94) number of shoots in wedge grafting during 15th-21st of July which was at par with (4.93) shoots in patch budding while, it was found minimum (4.75) in air layering. Similarly, data recorded during 15th-21st of July showed maximum (5.40) number of shoots in wedge grafting which was followed by (5.39) in patch budding while, minimum (3.78) in air layering.

4.1.5.6 Number of Shoots per Plant at 180 DAP

The data pertaining to plant height at 180 DAP of guava cv. L-49 as influenced by methods of propagation are presented in Table 9. It is clear from the data that results showed significant effect on the number of shoots per plant in guava at 180 DAP.

4.1.5.6.1 Influence of Time and Methods of Propagation on Number of Shoots per Plant in Guava cv. L-49

Data presented in Table 9 on number of shoots at 180 DAP of guava cv. L-49 as influenced by time and methods of propagation showed maximum (7.23) average number of shoots in wedge grafting followed by (6.66) in patch budding while, it was found minimum (4.88) in air layering. Similarly, the maximum (6.54) average number of shoots was observed during 15th-21st of August which was followed by (6.02) during 15th-21st of September, while minimum (5.30) during 15th-21st of June.

4.1.5.6.2 Interaction Effect of Time and Methods of Propagation on Number of Shoots per Plant in Guava cv. L-49

The data presented in Table 9 on interaction effect of time and methods of propagation recorded maximum (8.97) number of shoots in wedge grafting during 15th-21st of August which was followed by (6.88) patch budding while, minimum (4.94) in cuttings. Similarly, during 15th-21st of September maximum (6.85) number of shoots was recorded in wedge grafting which was at par with (6.76) patch budding while, it was observed minimum (5.21) in air layering.

Data pertaining to interaction effect of time and methods of propagation revealed maximum (6.81) number of shoots in wedge grafting during 15th-21st of July which was at par with patch budding (4.71) while, minimum (5.10) in air layering. Similarly, data recorded during 15th-21st of June showed maximum (6.30) number of shoots in wedge grafting which was at par with patch budding (6.28) while, minimum (4.27) in air layering.

4.1.6 Number of leaves per plant.

The data pertaining to the effect of time and methods of propagation on number of leaves at 30, 60, 90, 120, 150, 180 days after propagation in plants are presented in Table 10.

4.1.6.1 Number of Leaves per Plant at 30 DAP

The data obtained after 30 days on the influence of time and method of propagation on number of leaves per plant of guava cv. L-49 showed non significant differences presented in Table 10.

4.1.6.2 Number of Leaves per Plant at 60 DAP

The data presented in Table 10 revealed that both time and methods of propagation had significant effect on number of leaves per plant of guava.

4.1.6.2.1 Influence of Time and Methods of Propagation on Number of Leaves per Plant in Guava cv. L-49

Data presented in Table 10 on average number of leaves as influenced by time and methods of propagation recorded maximum (8.12) average number of leaves in wedge grafting which was followed by (7.96) patch budding while, it was found minimum (7.85) in cuttings. After 60 DAP the maximum (9.04) average number of leaves was observed during 15th-21st of August which was followed by (8.31) during 15th-21st of September while, minimum (6.47) during 15th-21st of June.

4.1.6.2.2 Interaction Effect of Time and Methods of Propagation on Number of Leaves per Plant in Guava cv. L-49

The data presented in Table 10 on interaction effect of methods and time of propagation recorded maximum (9.19) number of leaves in wedge grafting during 15th-21st of August which was followed by patch budding (9.09) while, the minimum (8.88) in air layering. Similarly, during 15th-21st of September maximum (8.41) number of leaves was recorded in wedge grafting which was followed by (8.19) patch budding while, minimum (5.11) in cuttings.

Data pertaining to interaction effect of time and methods of propagation recorded maximum (8.27) number of leaves in wedge grafting during 15th-21st July which was followed by patch budding (8.06) while, it was recorded minimum (7.92) in air layering. Similarly, during 15th-21st of June maximum (6.62) number of leaves was recorded in wedge grafting which was at par with (6.52) patch budding while, minimum (6.33) in cutting.

4.1.6.3 Number of Leaves per Plant at 90 DAP

It is apparent from the results presented in Table 10 that significant differences in both factors *i.e.* time and methods of propagation was observed with respect to number of leaves per plant at 90 DAP.

Table 10: Influence of Time, Method of Propagation and their Interaction Effect on Number of Leaves per Plant of Guava cv. L-49

Time of Propagation	*Method of Propagation*														
	June 15th-21st	*July 15th-21st*	*Aug 15th-21st*	*Sep 15th-21st*	*Mean*	*June 15th-21st*	*July 15th-21st*	*Aug 15th-21st*	*Sep 15th-21st*	*Mean*	*June 15th-21st*	*July 15th-21st*	*Aug 15th-21st*	*Sep 15th-21st*	*Mean*
	After 30 Days					*After 60 Days*					*After 90 Days*				
Cutting	1.24	2.32	3.56	2.47	2.40	6.33	7.94	9.01	8.11	7.85	10.22	11.25	12.10	12.87	**11.61**
Air Layering	1.83	3.27	3.33	4.33	3.19	6.42	7.92	8.88	8.52	7.93	10.13	11.80	11.95	11.87	**11.44**
Patch budding	1.83	1.90	1.81	1.26	1.70	6.52	8.06	9.09	8.19	7.96	11.39	12.48	14.12	13.56	**13.11**
Wedge grafting	1.53	2.15	2.73	2.55	2.24	6.62	8.27	9.19	8.41	8.12	12.22	13.68	15.00	14.11	**13.88**
Mean	1.61	2.41	2.86	2.65		6.47	8.06	9.04	8.31		10.99	12.30	13.65	13.10	

Factors	*S.E. m (±)*	*C.D (0.05)*	*S.E. m (±)*	*C.D (0.05)*	*S.E. m (±)*	*C.D (0.05)*
Method of propagation		N.S	0.03	0.09	0.01	0.03
Time of propagation		N.S	0.03	0.09	0.01	0.03
Method x Time		**N.S**	**0.06**	**0.18**	**0.02**	**0.07**

Contd...

Table 10–Contd...

Time of Propagation	*Method of Propagation*														
	June 15th-21st	*July 15th-21st*	*Aug 15th-21st*	*Sep 15th-21st*	*Mean*	*June 15th-21st*	*July 15th-21st*	*Aug 15th-21st*	*Sep 15th-21st*	*Mean*	*June 15th-21st*	*July 15th-21st*	*Aug 15th-21st*	*Sep 15th-21st*	*Mean*
	After 120 Days					*After 150 Days*					*After 180 days*				
Cutting	11.55	11.97	15.17	12.98	12.92	13.67	15.33	18.78	16.33	16.03	16.97	16.49	19.49	16.63	**16.89**
Air Layering	10.30	11.50	12.77	12.90	11.87	11.22	12.11	16.12	13.11	13.14	14.78	15.09	16.58	16.53	**15.74**
Patch budding	13.22	14.67	17.62	15.62	15.28	14.67	15.22	15.44	16.22	15.39	16.10	19.11	19.77	18.75	**18.43**
Wedge grafting	13.87	15.02	18.67	16.02	15.89	15.92	18.11	19.44	16.56	17.51	17.93	19.03	22.39	19.17	**19.63**
Mean	11.99	12.93	16.02	14.07		13.87	15.19	17.45	15.55		15.94	17.44	19.55	17.77	

Factors	*S.E. m (±)*	*C.D (0.05)*	*S.E. m (±)*	*C.D (0.05)*	*S.E. m (±)*	*C.D (0.05)*
Method of propagation	0.04	0.13	0.15	0.43	1.00	0.34
Time of propagation	0.04	0.13	0.15	0.43	1.00	0.34
Method x Time	**0.09**	**0.26**	**0.29**	**0.85**	**1.99**	**0.69**

4.1.6.3.1 Influence of Time and Methods of Propagation on Number of Leaves per Plant in Guava cv. L-49

Data presented in Table 10 on average number of leaves as influenced by time and methods of propagation recorded maximum (13.88) average number of leaves in wedge grafting which was followed by (13.11) patch budding while, minimum (11.44) in air layering. Similarly, during 15th-21st of August the number of leaves (13.65) was observed maximum which was followed by (13.10) during 15th-21st of September while, it was found minimum (10.99) during 15th-21st of June.

4.1.6.3.2 Interaction Effect of Time and Method of Propagation on Number of Leaves per Plant in Guava cv. L-49

The data presented in Table 10 on interaction effect of time and methods of propagation showed maximum (15.00) number of leaves in wedge grafting during 15th-21st of August which was followed by (14.12) patch budding while, minimum (11.95) in air layering. Similarly, during 15th-21st of September maximum (14.11) number of leaves was recorded in wedge grafting which was followed by patch budding (13.56) while, minimum (11.87) in layering.

Data pertaining to interaction effect of time and methods of propagation revealed maximum (13.68) number of leaves in wedge grafting during 15th-21st of July which was followed by patch budding (12.48) while, minimum (11.25) in cuttings. Similarly, during 15th-21st of June maximum (12.22) number of leaves was recorded in wedge grafting which was followed by patch budding (11.39) while, minimum (10.13) in air layering.

4.1.6.4 Number of Leaves per Plant at 120 DAP

The data pertaining to number of leaves at 120 DAP of guava cv. L-49 as influenced by time and methods of propagation are presented in Table 10. It is clear from the data that results showed significant effect on number of leaves at 120 DAP.

4.1.6.4.1 Influence of Time and Methods of Propagation on Number of Leaves per Plant in Guava cv. L-49

Data presented in Table 10 on average number of leaves at 120 DAP as influenced by time and methods of propagation showed maximum (15.89) average number of leaves in wedge grafting which was followed by (15.28) patch budding while, minimum (11.87) in air layering. Similarly, the maximum (16.02) average number of leaves was observed in plants propagated during 15th-21st of August which was followed by (14.07) during 15th-21st of September, while, minimum (11.99) during 15th-21st of June.

4.1.6.4.2 Interaction Effect of Time and Methods of Propagation on Number of Leaves per Plant in Guava cv. L-49

The data presented in Table 10 on interaction effect of time and methods of propagation showed that maximum (18.67) number of leaves was recorded in wedge grafting during 15th-21st of August which was followed by patch budding (17.62) while, minimum (12.77) in air layering. Similarly, during 15th-21st of September maximum (16.02) number of leaves was recorded in wedge grafting which was followed by patch budding (15.62) while, minimum (12.90) air layering.

Data pertaining to interaction effect of time and methods of propagation recorded maximum (15.02) number of leaves in wedge grafting during 15th-21st of July which was followed by patch budding (11.50) while, it was found minimum (11.50) in air layering. Similarly, during 15th-21st of June maximum (13.87) number of leaves was recorded in wedge grafting which was followed by (13.22) patch budding while, minimum (10.30) in air layering.

4.1.6.5 Number of Leaves per Plant at 150 DAP

The data pertaining to number of leaves at 150 DAP of guava cv. L-49 as influenced by methods and time is presented in Table 10. It is clear from the data that results showed significant effect on number of leaves at 150 DAP.

4.1.6.5.1 Influence of Time and Methods of Propagation on Number of Leaves per Plant in Guava cv. L-49

Data presented in Table 10 on average number of leaves at 150 DAP as influenced by time and methods of propagation showed maximum (17.51) average number of leaves in wedge grafting which was followed by (15.39) patch budding while, minimum (13.14) in air layering. Similarly, the maximum (17.45) average number of leaves was observed during 15th-21st of August which was followed by (15.55) during 15th-21st of September while, minimum (13.87) during 15th-21st of June.

4.1.6.5.2 Interaction Effect of Time and Methods of Propagation on Number of Leaves per Plant in Guava cv. L-49

The data presented in Table 10 on interaction effect of time and methods of propagation recorded maximum (19.44) number of leaves in wedge grafting during 15th-21st of August which was followed by patch budding (15.44) while, minimum (16.12) in air layering. Similarly, during 15th-21st of September maximum (16.56) number of leaves was recorded in wedge grafting which was followed by patch budding (16.22) while, minimum (13.11) in air layering.

Data pertaining to interaction effect of time and methods of propagation revealed maximum (18.11) number of leaves in wedge grafting during 15th-21st of July which was followed by (15.22) patch budding while, it was found minimum (12.11) in air layering. Similarly, during 15th-21st of June showed maximum (15.92) number of leaves was recorded in wedge grafting which was followed by patch budding (14.67) while, minimum (11.22) in air layering.

4.1.6.6 Number of Leaves per Plant at 180 DAP

The results related to effect of time and methods of propagation of guava plants, with respect to number of leaves per plant at the end of observation period (180 DAP) is presented in Table 10.

4.1.6.6.1 Influence of Time and Methods of Propagation on Number of Leaves per Plant in Guava cv. L-49

Data presented in Table 10 on number of leaves at 180 DAP as influenced by time and methods of propagation showed maximum (19.63) average number of leaves in wedge grafting which was followed by (18.43) patch budding while, minimum (15.74) in air layering. Similarly, the maximum (19.55) average number of leaves was recorded during 15th-21st of August which was followed by (17.77) during 15th-21st of September while, minimum (15.94) during 15th-21st of June.

4.1.6.6.2 Interaction Effect of Time and Methods of Propagation on Number of Leaves per Plant in Guava cv. L-49

The data presented in Table 10 on interaction effect of time and methods of propagation recorded maximum (22.39) number of leaves in wedge grafting during 15th-21st of August which was followed by (19.77) patch budding while, minimum (16.58) in air layering. Similarly, during 15th-21st of September maximum (19.17) number of leaves was observed in wedge grafting which was followed by (18.75) patch budding while, minimum (16.53) in air layering.

Data pertaining to interaction effect of time and methods of propagation recorded maximum (19.11) number of leaves in patch budding during 15th-21st of July which was followed by (19.03) wedge grafting while, minimum (15.09) in air layering. Similarly, during 15th-21st of June maximum (17.93) number of leaves was observed in wedge grafting which was followed by (16.10) patch budding while, minimum (14.78) in air layering.

4.1.7 Stem Thickness of Cutting/Layered/Budded/Grafted Plants

The data pertaining to the effect of time and methods of propagation on stem thickness at 30, 60, 90, 120, 150, 180 days after propagation are presented in Table11.

4.1.7.1 Stem Thickness at 30 DAP (cm)

The data obtained after 30 days on the influence of time and method of propagation on stem thickness of guava cv. L-49 showed non significant differences presented in Table 11.

4.1.7.2 Stem Thickness at 60 DAP (cm)

The data presented in Table 11 revealed that both time and methods of propagation had significant effect on stem thickness of guava.

4.1.7.2.1 Influence of Time and Methods of Propagation on Stem Thickness (cm) in Guava cv. L-49

Data presented in Table 11 related to average stem thickness as influenced by time and methods of propagation recorded maximum (1.09 cm) average stem thickness in wedge grafting which was followed by (1.05 cm) patch budding while, it was found minimum (0.91 cm) in air layering. After 60 DAP, the maximum (1.09 cm) average stem thickness was recorded during 15th-21st of August which was

Table 11: Influence of Time, Method of Propagation and their Interaction Effect on Stem Thickness of Guava cv. L-49

Time of Propagation	*Method of Propagation*														
	June 15th-21st	*July 15th-21st*	*Aug 15th-21st*	*Sep 15th-21st*	*Mean*	*June 15th-21st*	*July 15th-21st*	*Aug 15th-21st*	*Sep 15th-21st*	*Mean*	*June 15th-21st*	*July 15th-21st*	*Aug 15th-21st*	*Sep 15th-21st*	*Mean*
	After 30 Days					*After 60 Days*					*After 90 Days*				
Cutting	0.79	0.88	1.07	1.02	0.94	0.85	1.06	1.08	1.07	1.01	0.97	1.10	1.11	1.09	**1.07**
Air Layering	0.77	0.87	0.93	0.88	0.87	0.79	0.89	1.00	0.96	0.91	0.86	1.00	1.07	1.06	**1.00**
Patch budding	0.66	0.65	0.79	0.68	0.70	0.88	1.09	1.14	1.12	1.05	0.99	1.11	1.17	1.14	**1.10**
Wedge grafting	0.76	0.80	0.91	0.82	0.82	0.90	1.13	1.16	1.15	1.09	1.02	1.18	1.19	1.16	**1.14**
Mean	0.75	0.80	0.93	0.85		0.85	1.05	1.09	1.08		0.96	1.10	1.13	1.11	

Factors	*S.E. m (±)*	*C.D (0.05)*	*S.E. m (±)*	*C.D (0.05)*	*S.E. m (±)*	*C.D (0.05)*
Method of propagation		N.S	0.01	0.02	0.01	0.01
Time of propagation		N.S	0.01	0.02	0.01	0.01
Method x Time		**N.S**	**0.01**	**0.05**	**0.01**	**0.03**

Contd...

Table 11–*Contd...*

Time of Propagation	*Method of Propagation*														
	June 15th-21st	*July 15th-21st*	*Aug 15th-21st*	*Sep 15th-21st*	*Mean*	*June 15th-21st*	*July 15th-21st*	*Aug 15th-21st*	*Sep 15th-21st*	*Mean*	*June 15th-21st*	*July 15th-21st*	*Aug 15th-21st*	*Sep 15th-21st*	*Mean*
	After 120 Days					*After 150 Days*					*After 180 days*				
Cutting	1.07	1.14	1.18	1.16	1.14	1.14	1.18	1.23	1.21	1.19	1.21	1.25	1.29	1.27	**1.25**
Air Layering	1.05	1.05	1.14	1.09	1.08	1.10	1.15	1.21	1.19	1.16	1.13	1.21	1.25	1.23	**1.20**
Patch budding	1.09	1.15	1.21	1.20	1.16	1.16	1.18	1.26	1.25	1.21	1.21	1.29	1.33	1.31	**1.28**
Wedge grafting	1.11	1.18	1.24	1.22	1.19	1.17	1.25	1.31	1.28	1.25	1.23	1.30	1.34	1.32	**1.30**
Mean	1.08	1.13	1.19	1.17		1.14	1.19	1.25	1.23		1.20	1.26	1.30	1.28	

Factors	*S.E. m (±)*	*C.D (0.05)*	*S.E. m (±)*	*C.D (0.05)*	*S.E. m (±)*	*C.D (0.05)*
Method of propagation	0.01	0.02	0.01	0.01	0.01	0.01
Time of propagation	0.01	0.02	0.01	0.01	0.01	0.01
Method x Time	**0.01**	**0.05**	**0.01**	**0.02**	**0.01**	**0.02**

followed by (1.08 cm) during 15th-21st of September while, minimum (0.85 cm) during 15th-21st of June.

4.1.7.2.2 Interaction Effect of Time and Methods of Propagation on Stem Thickness (cm) in Guava cv. L-49

The data presented in Table 11 on interaction effect of time and methods of propagation recorded maximum (1.16 cm) stem thickness in wedge grafting during 15th-21st of August which was at par with (1.14 cm) patch budding while, the minimum (1.00 cm) in air layering. Similarly, during 15th-21st of September maximum (1.15 cm) stem thickness was recorded in wedge grafting which was at par with (1.12 cm) patch budding while, minimum (0.96 cm) in air layering.

Data pertaining to interaction effect of time and methods of propagation recorded maximum (1.13 cm) stem thickness in wedge grafting during 15th-21st of July which was at par with (1.09 cm) patch budding while, minimum (0.89 cm) in air layering. Similarly, during 15th-21st of June maximum (0.90 cm) stem thickness was recorded in wedge grafting which was at par with (0.88 cm) patch budding and cuttings (0.85 cm) while, minimum (0.79 cm) in air layering.

4.1.7.3 Stem Thickness at 90 DAP (cm)

It is apparent from the results presented in Table 11 that significant difference in both factors *i.e.* time and methods of propagation was observed with respect to stem thickness at 90 DAP.

4.1.7.3.1 Influence of Time and Methods of Propagation on Stem Thickness (cm) in Guava cv. L-49

Data presented in Table 11 on stem thickness as influenced by time and methods of propagation recorded maximum (1.14 cm) average stem thickness in wedge grafting which was followed by (1.10 cm) patch budding while, minimum (1.00 cm) in air layering. Similarly, on 15th-21st of August, the stem thickness (1.13 cm) was observed maximum which was followed by (1.11 cm) during 15th-21st of September while, minimum (0.96 cm) during 15th-21st of June.

4.1.7.3.2 Interaction Effect of Time and Methods of Propagation on Stem Thickness (cm) in Guava cv. L-49

The data presented in Table 11 on interaction effect of time and methods of propagation recorded maximum (1.19 cm) stem thickness in wedge grafting during 15th-21st of August which was at par with (1.17 cm) patch budding while, minimum (1.07 cm) in air layering. Similarly, during 15th-21st of September maximum (1.16 cm) stem thickness was recorded in wedge grafting which was at par with (1.14 cm) patch budding while, minimum (1.06 cm) in layering.

Data pertaining to interaction effect of time and methods of propagation recorded maximum (1.18 cm) stem thickness in wedge grafting during 15th-21st of July which was followed by (1.11 cm) patch budding while, it was found minimum (1.00 cm) in air layering. Similarly, during 15th-21st of June maximum (1.02 cm) stem thickness was recorded in wedge grafting which was followed by (0.99 cm) patch budding while, minimum (0.97 cm) in air layering.

4.1.7.4 Stem Thickness at 120 DAP (cm)

The data pertaining to stem thickness at 120 DAP of guava cv. L-49 as influenced by time and methods of propagation are presented in Table 11. It is clear from the data that results showed significant effect on the stem thickness of guava at 120 DAP.

4.1.7.4.1 Influence of Time and Methods of Propagation on Stem Thickness (cm) in Guava cv. L-49

Data presented in Table 11 on average stem thickness at 120 DAP of guava cv. L-49 as influenced by time and methods of propagation showed maximum (1.19 cm) average stem thickness in wedge grafting which was followed by (1.16 cm) in patch budding while, minimum (1.08 cm) in air layering. Similarly, the maximum (1.19 cm) average stem thickness was observed during 15th-21st of August which was followed by (1.17 cm) during 15th-21st of September while, minimum (1.08 cm) during 15th-21st of June.

4.1.7.4.2 Interaction Effect of Time and Methods of Propagation on Stem Thickness (cm) in Guava cv. L-49

The data presented in Table 11 on interaction effect of time and methods of propagation recorded maximum (1.24 cm) stem thickness in wedge grafting during 15th-21st of August which was at par with (1.21 cm) patch budding while, minimum (1.14 cm) in air layering. Similarly, during 15th-21st of September maximum (1.22 cm) stem thickness was observed in wedge grafting which was at par with (1.20 cm) patch budding while, minimum (1.09 cm) in air layering.

Data pertaining to interaction effect of time and methods of propagation recorded maximum (1.18 cm) stem thickness in wedge grafting during 15th-21st of July which was at par with (1.15 cm) in patch budding while, it was found minimum (1.05 cm) in air layering. Similarly, during 15th-21st of June maximum (1.11 cm) stem thickness was observed in wedge grafting which was at par with (1.09 cm) patch budding while, minimum (1.05 cm) in air layering.

4.1.7.5 Stem Thickness at 150 DAP (cm)

The data pertaining to stem thickness at 150 DAP of guava cv. L-49 as influenced by time and methods of propagation are presented in Table 11. It is clear from the data that results showed significant effect on the plant height of guava at 150 DAP.

4.1.7.5.1 Influence of Time and Methods of Propagation on Stem Thickness (cm) in Guava cv. L-49

Data presented in Table 11 on average stem thickness at 150 DAP as influenced by time and methods of propagation showed maximum (1.25 cm) average stem thickness in wedge grafting which was followed by (1.21 cm) patch budding while, minimum (1.16 cm) in air layering. Similarly, the maximum (1.25 cm) average stem thickness was recorded in plants propagated during 15th-21st of August which was followed by (1.23 cm) during 15th-21st of September while, minimum (1.14 cm) during 15th-21st of June.

4.1.7.5.2 Interaction Effect of Time and Methods of Propagation on Stem Thickness (cm) in Guava cv. L-49

The data presented in Table 11 on interaction effect of time and methods of propagation recorded maximum (1.31 cm) stem thickness in wedge grafting during 15th-21st of August which was followed by (1.26 cm) patch budding while, minimum (1.21 cm) in air layering. Similarly, data obtained from different propagation during 15th-21st of September showed maximum (1.28 cm) stem thickness in wedge grafting which was followed by (1.25 cm) patch budding while, minimum (1.19 cm) in air layering.

Data pertaining to interaction effect of time and methods of propagation during 15th-21st of July recorded maximum (1.25 cm) stem thickness in wedge grafting which was followed by (1.18 cm) patch budding while, minimum (1.15 cm) in air layering. Similarly, during 15th-21st of June maximum (1.17 cm) stem thickness was recorded in wedge grafting which was at par with (1.16 cm) patch budding while, minimum (1.10 cm) in air layering.

4.1.7.6 Stem Thickness at 180 DAP (cm)

The data pertaining to stem thickness at 180 DAP of guava cv. L-49 as influenced by time and methods of propagation are presented in Table 11. It is clear from the data that results showed significant effect on the plant height of guava at 180 DAP.

4.1.7.6.1 Influence of Time and Methods of Propagation on Stem Thickness (cm) in Guava cv. L-49

Data presented in Table 11 on stem thickness at 180 DAP as influenced by time and methods of propagation showed maximum (1.30 cm) average stem thickness in wedge grafting which was followed by (1.28 cm) patch budding while, minimum (1.20 cm) in air layering. Similarly, the maximum (1.30 cm) average stem thickness was observed during 15th-21st of August which was followed by (1.28 cm) during 15th-21st of September while, minimum (1.20 cm) during 15th-21st of June.

4.1.7.6.2 Influence of Time and Methods of Propagation on Stem Thickness (cm) in Guava cv. L-49

The data presented in Table 11 on interaction effect of time and methods of propagation recorded maximum (1.34 cm) stem thickness in wedge grafting during 15th-21st of August which was at par with (1.33 cm) patch budding while, it was found minimum (1.25 cm) in air layering. Similarly, during 15th-21st of September maximum (1.32 cm) stem thickness was recorded in wedge grafting which was at par with (1.31 cm) patch budding while, minimum (1.23 cm) in air layering.

Data pertaining to interaction effect of time and methods of propagation showed maximum (1.30 cm) stem thickness in wedge grafting during 15th-21st of July which was at par with (1.29 cm) patch budding while, it was found minimum (1.21 cm) in air layering. Similarly, during 15th-21st of June maximum (1.23 cm) stem thickness was recorded in wedge grafting which was at par with (1.21 cm) patch budding and cuttings (1.21 cm) while, minimum (1.13 cm) in air layering.

4.1.8 Average Leaf Area (cm^2)

It is apparent from the results presented in Table 12 that significant difference in both factors *i.e.* time and methods of propagation was observed with respect to average leaf (cm^2) at 90 DAP.

4.1.8.1 Influence of Time and Methods of Propagation on Average Leaf Area (cm^2) in Guava cv. L-49

Data presented in Table 12 on average leaf area among different time and methods of propagation recorded maximum (15.47 cm^2) average leaf area in patch budding which was followed by (15.37 cm^2) wedge grafting while, minimum (15.34 cm^2) in air layering. At 90 DAP, maximum (18.20 cm^2) average leaf area was recorded during 15th-21st of August which was followed by (16.48 cm^2) during 15th-21st of September while, minimum (11.39 cm^2) during 15th-21st of June.

Table 12: Influence of Time, Method of Propagation and their Interaction Effect Average Leaf Area (cm^2) of Guava cv. L-49 after 90 Days

Method of Propagation	*Time of Propagation*				
	June 15th-21st	*July 15th-21st*	*Aug 15th-21st*	*Sep 15th-21st*	*Mean*
Cutting	11.68	15.28	18.17	16.27	15.35
Air Layering	11.18	15.42	18.18	16.58	15.34
Patch budding	11.42	15.41	18.18	16.88	15.47
Wedge grafting	11.28	15.42	18.28	16.48	15.37
Mean	**11.39**	**15.38**	**18.20**	**16.48**	

Factors	*S.E. m (±)*	*C.D (0.05)*
Method of propagation	0.01	0.01
Time of propagation	0.01	0.01
Method x Time	**0.01**	**0.01**

4.1.8.1 Interaction Effect of Time and Methods of Propagation on Average Leaf Area (cm^2) in Guava cv. L-49

The data presented in Table 12 on interaction effect of time and methods of propagation recorded maximum (18.28 cm^2) leaf area in wedge grafting during 15th-21st of August which was followed by (11.18 cm^2) patch budding and air layering while, it was found minimum (18.17 cm^2) in cuttings. Similarly, during 15th-21st of September maximum (16.88 cm^2) leaf area was recorded in patch budding which was followed by (16.58 cm^2) air layering while, it was found minimum (16.27 cm^2) in cuttings.

Data pertaining to interaction effect of time and methods of propagation recorded maximum (15.42 cm^2) leaf area in wedge grafting and air layering during 15th-21st of July while, minimum (15.28 cm^2) in cuttings. Similarly, during 15th-21st of June maximum (11.68 cm^2) leaf area was recorded in cuttings which was followed by (11.28 cm^2) patch budding while, minimum (11.18 cm^2) in air layering.

4.1.9 Average Leaf Fresh Weight (g)

The data pertaining to average leaf fresh weight (g) at 90 DAP of guava cv. L-49 as influenced by time and methods of propagation are presented in Table 13. It is clear from the data that results showed significant effect on average leaf fresh weight of guava at 120 DAP.

Table 13: Influence of Time, Method of Propagation and their Interaction Effect on Average Leaf Fresh Weight (g) of Guava cv. L-49 after 90 Days

Method of Propagation	*Time of Propagation*				
	June 15th-21st	*July 15th-21st*	*Aug 15th-21st*	*Sep 15th-21st*	*Mean*
Cutting	0.22	0.23	0.32	0.26	0.26
Air Layering	0.20	0.21	0.30	0.26	0.24
Patch budding	0.28	0.26	0.32	0.30	0.29
Wedge grafting	0.29	0.27	0.34	0.31	0.30
Mean	0.24	0.24	0.32	0.28	

Factors	*S.E. m (±)*	*C.D (0.05)*
Method of propagation	0.01	0.01
Time of propagation	0.01	0.01
Method x Time	**0.01**	**0.01**

4.1.9.1 Influence of Time and Methods of Propagation on Average Leaf Fresh Weight (g) in Guava cv. L-49

Data presented in Table 13 showed maximum (0.30 g) average leaf fresh weight in wedge grafting which was at par with (0.29 g) patch budding while, it was found minimum (0.24 g) in air layering. Similarly, maximum (0.32 g) average leaf fresh weight was observed during 15th-21st of August which was followed by (0.28 g) during 15th-21st of September while, minimum (0.24 g) during 15th-21st of June.

4.1.9.2 Interaction Effect of Time and Methods of Propagation on Average Leaf Fresh Weight (g) in Guava cv. L-49

The data presented in Table 13 on effect of time and methods of propagation recorded maximum (0.34 g) leaf fresh weight in wedge grafting during 15th-21st of August which was followed by (0.32 g) patch budding while, it was recorded minimum (0.30 g) in air layering. Similarly, data obtained from different propagation methods during 15th-21st of September showed maximum (0.31 g) leaf fresh weight in wedge grafting which was at par with (0.30 g) patch budding while, minimum (0.26 g) in air layering and cuttings.

Data pertaining to interaction effect of time and methods of propagation recorded maximum (0.27 g) leaf fresh weight in wedge grafting during 15th-21st of July which was at par with (0.26 g) patch budding while, it was recorded minimum (0.21 g) in air layering. Similarly, during 15th-21st of June maximum (0.29 g) leaf fresh weight was recorded in wedge grafting which was at par with (0.29 g) patch budding while, minimum (0.20 g) in air layering.

4.1.10 Average Leaf Dry Weight (mg)

It is apparent from the results presented in Table 14 indicated that significant differences in both time and methods of propagation was observed with respect to average leaf dry weight (g) at 90 DAP.

Table 14: Influence of Time, Method of Propagation and their Interaction Effect on Average Leaf Dry Weight (mg) of Guava cv. L-49 after 90 Days

Method of Propagation	*Time of Propagation*				
	June 15th-21st	*July 15th-21st*	*Aug 15th-21st*	*Sep 15th-21st*	*Mean*
Cutting	16.65	20.00	23.63	21.66	20.50
Air Layering	16.00	17.17	22.32	19.90	18.85
Patch budding	19.90	21.27	24.10	23.13	22.10
Wedge grafting	21.12	23.23	25.96	25.07	23.85
Mean	**18.42**	**20.42**	**24.02**	**22.44**	

Factors	*S.E. m (±)*	*C.D (0.05)*
Method of propagation	0.12	0.35
Time of propagation	0.12	0.35
Method x Time	**0.24**	**0.69**

4.1.10.1 Influence of Time and Methods of Propagation on Average Leaf Dry Weight (mg) in Guava cv. L-49

Data presented in Table 14 as influenced by time and methods of propagation showed maximum (23.85 mg) average leaf dry weight in wedge grafting which was followed by patch budding (22.10 mg) while, minimum (18.85 mg) in air layering. Similarly, maximum (24.02 mg) average leaf dry weight was observed during 15th-21st of August which was followed by (22.44 mg) during 15th-21st of September while, minimum (18.42 mg) during 15th-21st of June.

4.1.10.2 Interaction Effect of Time and Methods of Propagation on Average Leaf Dry Weight (mg) in Guava cv. L-49

The data presented in Table 14 on interaction effect of methods and time of propagation revealed maximum (25.96 mg) leaf dry weight in wedge grafting during 15th-21st of August which was followed by (24.10 mg) patch budding while, it was found minimum (22.32 mg) in air layering. Similarly, data obtained from different propagation methods during 15th-21st of September showed maximum (25.07 mg) leaf dry weight in patch budding which was followed by (23.13 mg) patch budding while, minimum (19.90 mg) in cuttings.

Data pertaining to interaction effect of time and methods of propagation recorded maximum (23.23 mg) leaf dry weight in wedge grafting during 15th-21st of

July which was followed by (21.29 mg) patch budding while, it was found minimum (17.17 mg) in air layering. Similarly, during 15th-21st of June maximum (21.12 mg) leaf dry weight was recorded in wedge grafting which was followed by (19.90 mg) patch budding while, minimum (16.00 mg) in air layering.

4.1.11 Chlorophyll Percentage (after 90 days)

The data pertaining to chlorophyll percentage at 90 DAP of guava cv. L-49 as influenced by time and methods of propagation are presented in Table 15. It is clear from the data that results showed significant effect on chlorophyll percentage of guava.

Table 15: Influence of Time, Method of Propagation and their Interaction Effect on Chlorophyll Percentage of Guava cv. L-49 after 90 Days

Method of Propagation	*Time of Propagation*				
	June 15th-21st	*July 15th-21st*	*Aug 15th-21st*	*Sep 15th-21st*	*Mean*
Cutting	1.67 (0.11)	1.99 (0.14)	2.28 (0.17)	2.16 (0.15)	2.02 (0.14)
Air Layering	1.65 (0.10)	1.94 (0.13)	2.23 (0.16)	2.11 (0.14)	1.98 (0.13)
Patch budding	2.03 (0.15)	1.94 (0.15)	2.28 (0.17)	2.21 (0.16)	2.12 (0.15)
Wedge grafting	2.12 (0.14)	1.91 (0.15)	2.46 (0.19)	2.14 (0.15)	2.16 (0.15)
Mean	**1.87 (0.12)**	**1.94 (0.13)**	**2.31 (0.17)**	**2.16 (0.15)**	
Factors		*S.E. m (±)*		*C.D (0.05)*	
Method of propagation		0.01		0.01	
Time of propagation		0.01		0.01	
Method x Time		**0.01**		**0.02**	

Note: Figures in parenthesis indicate observed values and others are transformed values.

4.1.11.1 Influence of Time and Methods of Propagation on Chlorophyll Percentage in Guava cv. L-49

Data presented in Table 15 as influenced by time and methods of propagation showed the maximum (0.15 per cent) average chlorophyll percentage in wedge grafting and patch budding which was par with (0.14 per cent) cuttings while, it was found minimum (0.13 per cent) in air layering. Similarly, the maximum (0.17 per cent) average chlorophyll percentage was observed during 15th-21st of August, which was followed by (0.15 per cent) during 15th-21st of September, while minimum (0.12 per cent) during 15th-21st of June.

4.1.11.2 Interaction Effect of Time and Methods of Propagation on Chlorophyll Percentage in Guava cv. L-49

The data presented in Table 15 on interaction effect of methods and time of propagation recorded maximum (0.19 per cent) chlorophyll percentage in wedge grafting during 15th-21st of August which was followed by (0.17 per cent) patch budding and cuttings while, minimum (0.16 per cent) in air layering. Similarly, during 15th-21st of September maximum (0.16 per cent) chlorophyll percentage was recorded in wedge grafting which was followed by (0.15 per cent) patch budding and cutting while, minimum (0.14 per cent) in air layering.

Data pertaining to interaction effect of time and methods of propagation recorded maximum (0.15 per cent) chlorophyll percentage in wedge grafting and patch budding during 15th-21st of July which was followed by (0.14) cuttings while, minimum (0.13 per cent) in air layering. Similarly, during 15th-21st of June maximum (0.15 per cent) chlorophyll percentage was recorded in patch budding which was at par with (0.14 per cent) wedge grafting while, minimum (0.10 per cent) in air layering.

4.2 Experiment II

To find out best soil media and time for guava propagation through cuttings under Jammu sub-tropics.

4.2.1 Percentage of Rooted Cutting (after 90 days of planting)

The data pertaining to influence of time and soil media for guava propagation through cuttings on percentage of rooted cuttings after 90 days presented in Table 16 showed significant differences among all the treatments.

4.2.1.1 Influence of Time and Media for Planting Cuttings on Percentage of Rooted Cuttings in Guava cv. L-49

The data presented in Table 16 showed highest (70.37 per cent) average percentage of rooted cuttings after 90 days in treatment combination vermiculite + sand + FYM (1:1:1) which was followed by (69.04 per cent) in Vermiculite + sand (1:1) while, it was found lowest (42.31 per cent) in sand + FYM (1:1). Similarly, highest (69.98 per cent) average percentage of rooted cuttings was recorded during 15th-21st of August which was followed by (62.06 per cent) during 15th-21st of September while, lowest (50.80 per cent) during 15th-21st of June.

4.2.1.2 Interaction Effect of Time and Media for Planting Cuttings on Percentage of Rooted Cuttings in Guava cv. L-49

The data presented in Table 16 on interaction effect of time and soil media for guava propagation through cuttings recorded highest (78.69 per cent) percentage of rooted cuttings during 15th-21st of August in the treatment comprising Vermiculite + sand + FYM (1:1:1) which was at par with (77.02 per cent) in Vermiculite + sand (1:1) and (76.01 per cent) in perlite + sand + FYM (1:1:1) while, lowest (48.77 per cent) in sand + FYM (1:1). Similarly, during 15th-21st of September highest (71.01 per cent) percentage of rooted cuttings was recorded in Vermiculite + sand + FYM (1:1:1) which was at par with (69.04 per cent) in Vermiculite + sand (1:1) followed by (65.81 per cent) in perlite + sand + FYM while, it was found lowest (47.05 per cent) in sand + FYM (1:1).

Data pertaining to interaction effect of time and soil media for guava propagation through cuttings during 15th-21st of July recorded highest (69.71 per cent) percentage of rooted cuttings in treatment comprising Vermiculite + sand + FYM in ratio 1:1:1

which was at par with (68.70 per cent) in Vermiculite + sand (1:1) and (67.01 per cent) in perlite + sand + FYM (1:1:1) while it was recorded lowest (41.68 per cent) in sand + FYM (1:1). Similarly, data obtained from planting cuttings during 15th-21st of June recorded highest (62.09 per cent) percentage of rooted cuttings in soil media comprising Vermiculite + sand + FYM (1:1:1) which was at par with (61.41 per cent) in Vermiculite + sand (1:1) and (60.20 per cent) in perlite + sand + FYM (1:1:1) while, lowest (31.74 per cent) percentage of rooted cuttings was observed in sand + FYM (1:1).

Table 16: Influence of Soil Media, Time of Planting Cuttings and their Interaction Effect on Percentage of Rooted Cuttings of Guava cv. L-49 after 90 Days

Method of Propagation	*Cuttings Soil Media*				
	June 15th-21st	*July 15th-21st*	*Aug 15th-21st*	*Sep 15th-21st*	*Mean*
Vermiculite + Sand	51.70 (61.41)	56.37 (68.70)	61.85 (77.02)	56.30 (69.04)	56.49 (69.04)
Perlite + Sand	38.42 (38.73)	50.92 (60.23)	56.66 (69.41)	49.29 (57.41)	49.29 (56.45)
Sand + FYM	34.18 (31.74)	50.92 (41.68)	44.28 (48.77)	49.29 (47.05)	40.48 (42.31)
Vermiculite + Sand + FYM	52.05 (62.09)	56.71 (69.71)	62.75 (78.69)	57.60 (71.01)	57.28 (70.37)
Perlite + Sand+ FYM	50.95 (60.20)	55.05 (67.01)	61.31 (76.01)	54.31 (65.81)	55.41 (67.22)
Mean	**45.46 (50.80)**	**51.79 (61.47)**	**57.37 (69.98)**	**52.16 (62.06)**	

Factors	*S.E. m (±)*	*C.D (0.05)*
Soil media	0.56	1.60
Time of planting cuttings	0.56	1.43
Media x Time	**1.12**	**3.20**

Note: Figures in parenthesis indicate observed values and others are transformed values.

4.2.2 Number of Days taken to Sprout

The data pertaining to influence of time and soil media for guava propagation through cuttings on number of days taken to sprout presented in Table 17 showed significant differences among all the treatments.

4.2.2.1 Influence of Time and Media for Planting Cuttings on Number of Days to Sprout in Guava cv. L-49

The data presented in Table 17 showed that minimum (9.96) average number of days to sprout by cuttings was observed in treatment combination of perlite + sand + FYM (1:1:1) which was followed by (10.13) in perlite + sand (1:1) while, maximum (18.97) in sand + FYM (1:1). Similarly, during 15th-21st of August minimum (12.46) average number of days to sprout was observed which was followed by (13.56) during 15th-21st of September while, maximum (14.19) during 15th-21st of June.

4.2.2.2 Interaction Effect of Time and Media of Planting Cuttings on Number of Days to Sprout in Guava cv. L-49 after 90 Days

The data presented in Table 17 on interaction effect of time and soil media for guava propagation through cuttings recorded minimum (8.95) number of days to sprout during 15th-21st of August in treatment combination perlite+ sand + FYM (1:1:1) which was followed by (9.56) in perlite + sand (1:1) while, maximum (16.87 days) in perlite + FYM (1:1). Similarly, during 15th-21st of September minimum (9.94) number of days taken to sprout was observed in perlite + sand + FYM (1:1:1) which was followed by (9.97) in perlite + sand (1:1) while, maximum (18.86) in sand + FYM (1:1).

Table 17: Influence of Soil Media, Time of Planting Cuttings and their Interaction Effect on Number of Days taken to Sprouting of Guava Cuttings cv. L-49

Method of Propagation	*Cuttings Soil Media*				
	June 15th-21st	*July 15th-21st*	*Aug 15th-21st*	*Sep 15th-21st*	*Mean*
Vermiculite + Sand	18.51	19.95	16.85	18.86	18.55
Perlite + Sand	10.85	10.16	9.56	9.97	10.13
Sand + FYM	20.18	19.96	16.87	18.86	18.97
Vermiculite + Sand + FYM	11.06	10.27	9.96	10.10	10.35
Perlite + Sand+ FYM	10.27	10.05	8.95	9.94	9.96
Mean	**14.19**	**14.09**	**12.46**	**13.56**	

Factors	*S.E. m (±)*	*C.D (0.05)*
Soil media	0.17	0.50
Time of planting cuttings	0.15	0.44
Media x Time	**0.35**	**0.99**

Data pertaining to interaction effect of time and soil media for guava propagation through cuttings on 15th-21st of July recorded minimum (10.05) number of days taken to sprout in treatment combination perlite + sand + FYM (1:1:1) which was followed by (10.16) in perlite + sand (1:1) while, maximum (19.96) in sand + FYM (1:1). Similarly, data obtained from planting cuttings during 15th-21st of June recorded minimum (10.27) number of days to sprout in perlite + sand + FYM (1:1:1) which was followed by (10.85) in perlite+ sand (1:1) while, maximum (20.18 days) in sand + FYM (1:1).

4.2.3 Number of Shoots per Plant

The data pertaining to the effect of time and soil media for guava propagation through cuttings on number of shoots per plant at 30, 60, 90, 120, 150 and 180 days after planting are presented in Table 18.

4.2.3.1 Number of Shoots per Plant at 30 DAP

The data obtained after 30 days on the influence of time and soil media for guava propagation through cuttings on number of shoots per plant non significant differences presented in Table 18.

4.2.3.2 Number of Shoots per Plant at 60 DAP

The data presented in Table 18 revealed that both time and soil media for guava propagation through planting cuttings had significant influence on the number of shoots per guava cuttings after 60 days.

4.2.3.2.1 Influence of Soil Media and Time of Planting Cuttings on Number of Shoots per Plant in Guava cv. L-49

The data presented in Table 18 showed maximum (4.48) average number of shoots in treatment combination vermiculite + sand + FYM (1:1:1) which was followed by (3.92) in Vermiculite + sand (1:1) while, minimum (3.52) in sand + FYM (1:1). At 60 days, the maximum (4.19) average number of shoot was observed during 15th-21st of August which was followed by (4.01) during 15th-21st of September while, minimum (3.50) during 15th-21st of June.

4.2.3.2.2 Interaction Effect of Time and Soil Media for Planting Cuttings on Number of Shoots per Plant in Guava cv. L-49

The interaction effect of time and soil media for planting cuttings recorded maximum (4.88) number of shoots in Vermiculite + sand + FYM (1:1:1) during 15th-21st of August which was followed by (4.21) in Vermiculite + sand (1:1) while, minimum (3.74) in sand + FYM (1:1). Similarly, data obtained during 15th-21st of September showed maximum (4.64) number of shoots in Vermiculite + sand + FYM (1:1:1) which was followed by (4.08) in Vermiculite + sand (1:1) while, minimum (3.58) in sand + FYM (1:1).

Data pertaining to interaction effect of time and soil media for planting cuttings during 15th-21st of July recorded maximum (4.28) number of shoots in Vermiculite + sand + FYM (1:1:1) which was followed by (3.90) in Vermiculite + sand (1:1) while, it was recorded minimum (3.44) in sand + FYM (1:1). Similarly, during 15th-21st of June maximum (4.12) number of shoots was recorded in Vermiculite + sand + FYM (1:1:1) which was followed (3.49) in Vermiculite + sand (1:1) while, minimum (3.31) in sand + FYM (1:1).

4.2.3.3 Number of Shoots per Plant at 90 DAP

The data pertaining to influence of time and soil media for guava propagation through cuttings on number of shoots after 90 days presented in Table 18 showed significant differences among all the treatments.

4.2.3.3.1 Influence of Time and Soil Media for Planting Cuttings on Number of Shoots per Plant in Guava cv. L-49

The data presented in Table 18 on effect of time and soil media recorded maximum (5.16) average number of shoots in treatment combination vermiculite + sand + FYM (1:1:1) which was followed by (4.85) in Vermiculite + sand (1:1) while,

Table 18: Influence of Soil Media, Time of Planting Cuttings and their Interaction Effect on Number of Shoots per Plant of Guava cv. L-49

Time of Propagation	*Cuttings Soil Media*														
	June 15th-21st	*July 15th-21st*	*Aug 15th-21st*	*Sep 15th-21st*	*Mean*	*June 15th-21st*	*July 15th-21st*	*Aug 15th-21st*	*Sep 15th-21st*	*Mean*	*June 15th-21st*	*July 15th-21st*	*Aug 15th-21st*	*Sep 15th-21st*	*Mean*
	After 30 Days					*After 60 Days*					*After 90 Days*				
Vermiculite + Sand	2.16	2.14	2.52	2.44	2.31	3.49	3.90	4.21	4.08	3.92	4.61	4.76	5.24	4.80	4.85
Perlite + Sand	1.95	2.02	2.13	2.15	2.06	3.19	3.72	4.00	3.85	3.69	3.91	4.00	4.41	4.16	4.12
Sand + FYM	2.06	1.89	2.02	2.05	2.00	3.31	3.44	3.74	3.58	3.52	3.70	3.80	3.99	3.90	3.85
Vermiculite + Sand + FYM	2.11	2.45	2.55	2.46	2.39	4.12	4.28	4.88	4.64	4.48	4.94	5.14	5.34	5.24	5.16
Perlite + Sand+ FYM	2.06	2.05	2.46	2.16	2.18	3.39	3.77	4.12	3.88	3.79	4.13	4.22	4.82	4.77	4.49
Mean	2.07	2.11	2.34	2.25		3.50	3.82	4.19	4.01		4.26	4.38	4.76	4.57	

Factors	*S.E. m (±)*	*C.D (0.05)*	*S.E. m (±)*	*C.D (0.05)*	*S.E. m (±)*	*C.D (0.05)*
Soil media		N.S	0.01	0.02	0.01	0.01
Time of planting cuttings		N.S	0.01	0.02	0.01	0.01
Media x Time		**N.S**	**0.01**	**0.04**	**0.01**	**0.03**

Contd...

Table 18–*Contd...*

Time of Propagation	*Cuttings Soil Media*														
	June 15th-21st	*July 15th-21st*	*Aug 15th-21st*	*Sep 15th-21st*	*Mean*	*June 15th-21st*	*July 15th-21st*	*Aug 15th-21st*	*Sep 15th-21st*	*Mean*	*June 15th-21st*	*July 15th-21st*	*Aug 15th-21st*	*Sep 15th-21st*	*Mean*
	After 120 Days					*After 150 Days*					*After 180 Days*				
Vermiculite +Sand	6.04	6.08	7.74	7.58	6.86	7.99	8.08	8.49	8.08	8.22	8.58	9.59	10.10	10.00	9.57
Perlite + Sand	4.90	5.93	5.68	5.59	5.53	6.58	6.98	7.12	6.98	6.94	8.09	8.28	9.29	9.14	8.70
Sand + FYM	4.58	5.00	4.87	4.80	4.82	5.88	5.99	6.78	5.99	6.31	6.66	7.11	7.30	7.21	7.07
Vermiculite + Sand + FYM	7.30	7.88	7.80	7.61	7.65	8.37	8.58	8.97	8.58	8.70	8.67	10.00	10.42	10.19	9.82
Perlite + Sand+ FYM	5.18	6.07	5.72	5.65	5.66	7.30	7.59	8.08	7.59	7.66	8.39	8.75	9.58	9.27	9.00
Mean	5.60	6.19	6.37	6.25		7.23	7.45	7.89	7.45		8.08	8.74	9.34	9.16	

Factors	*S.E. m (±)*	*C.D (0.05)*	*S.E. m (±)*	*C.D (0.05)*	*S.E. m (±)*	*C.D (0.05)*
Soil media	0.01	0.01	0.01	0.04	0.04	0.10
Time of planting cuttings	0.01	0.01	0.01	0.04	0.04	0.10
Media x Time	**0.01**	**0.02**	**0.03**	**0.09**	**0.07**	**0.22**

minimum (3.85) in sand + FYM (1:1). After 90 days, the maximum (4.76) average number of shoot was observed during 15th-21st of August which was followed by (4.57) during 15th-21st of September while, minimum (4.26) during 15th-21st of June.

4.2.3.3.2 Interaction Effect of Time and Media for Planting Cuttings on Number of Shoots per Plant in Guava cv. L-49

The data presented in Table 18 on interaction effect of time and soil media for guava propagation through cuttings recorded (5.34) maximum number of shoots in treatment combination Vermiculite + sand + FYM (1:1:1) during 15th-21st of August which was followed by (5.24) in Vermiculite + sand (1:1) while, minimum (3.99) in sand + FYM (1:1). Similarly, data obtained during 15th-21st of September showed maximum (5.24) number of shoots in Vermiculite + sand + FYM (1:1:1) which was followed by (4.80) in Vermiculite + sand (1:1) while, minimum (3.90) in sand + FYM (1:1).

Data pertaining to interaction effect of time and soil media for guava propagation through cuttings during 15th-21st of July recorded maximum (5.14) number of shoots in treatment combination Vermiculite + sand + FYM (1:1:1) which was followed by (4.76) in Vermiculite + sand (1:1) while, minimum (3.80) in sand + FYM (1:1). Similarly, during 15th-21st of June maximum (4.94) number of shoots was recorded in Vermiculite + sand + FYM (1:1:1) which was followed by (4.61) in Vermiculite + sand (1:1) while, minimum (3.70) in sand + FYM (1:1).

4.2.3.4 Number of Shoots per Plant at 120 DAP

The data pertaining to influence of time and soil media for guava propagation through cuttings on number of shoots per plant after 120 days presented in Table 18 showed significant differences among all the treatments.

4.2.3.4.1 Influence of Time and Soil Media for Planting Cuttings on Number of Shoots per Plant in Guava cv. L-49

The data presented in Table 18 showed that maximum (7.65) average number of shoots was recorded in treatment combination vermiculite + sand + FYM (1:1:1) which was followed by (6.86) in Vermiculite + sand (1:1) while, minimum (4.82) in sand + FYM (1:1). After 120 days, the maximum (6.37) average number of shoot was observed during 15th-21st of August which was followed by (6.25) during 15th-21st of September while, minimum (5.60) during 15th-21st of June.

4.2.3.4.2 Interaction Effect of Time and Soil Media for Planting Cuttings on Number of Shoots per Plant in Guava cv. L-49

The data presented in Table 18 on interaction effect of time and soil media for guava propagation through cuttings recorded maximum (7.80) number of shoots in Vermiculite + sand + FYM (1:1:1) during 15th-21st of August which was followed by (7.74) in Vermiculite + sand (1:1) while, minimum (4.87) in sand + FYM (1:1). Similarly, data obtained during 15th-21st of September showed maximum (7.61) number of shoots in Vermiculite + sand + FYM (1:1:1) which was followed by (7.58) in Vermiculite + sand (1:1) while, minimum (4.80) in sand + FYM (1:1).

Data pertaining to interaction effect of time and soil media for planting cuttings recorded maximum (7.88) number of shoots in Vermiculite + sand + FYM (1:1:1) during 15th-21st of July which was followed by (6.08) in Vermiculite + sand (1:1) while, minimum (5.00) in sand + FYM (1:1). Similarly, during 15th-21st of June maximum (7.30) number of shoots was recorded in Vermiculite + sand + FYM (1:1:1) which was followed by (6.04) in Vermiculite + sand (1:1) while, minimum (4.58) in sand + FYM (1:1).

4.2.3.5 Number of Shoots per Plant at 150 DAP

The data pertaining to influence of time and soil media for guava propagation through cuttings on number of shoots after 150 days presented in Table 18 showed significant differences among all the treatments.

4.2.3.5.1 Influence of Time and Soil Media for Planting Cuttings on Number of Shoots per Plant in Guava cv. L-49

Data presented in Table 18 showed maximum (8.70) average number of shoots in treatment combination vermiculite + sand + FYM (1:1:1) which was followed by (8.22) in Vermiculite + sand (1:1) while, minimum (6.31) in sand + FYM (1:1). After 150 days, the maximum (7.89) average number of shoot was observed during 15th-21st of August which was followed by (7.45) during 15th-21st of September while, minimum (7.23) during 15th-21st of June.

4.2.3.5.2 Interaction Effect of Time and Soil Media for Planting Cuttings on Number of Shoots per Plant in Guava cv. L-49

The data presented in Table 18 on interaction effect of time and soil media for guava propagation through cuttings recorded maximum (8.97) number of shoots during 15th-21st of August in Vermiculite + sand + FYM (1:1:1) which was followed by (8.49) in Vermiculite + sand (1:1) while, average (6.78) in sand + FYM (1:1). Similarly, during 15th-21st of September maximum (8.58) number of shoots was recorded in Vermiculite + sand + FYM (1:1:1) which was followed by (8.08) in Vermiculite + sand (1:1) while, minimum (5.99) in sand + FYM (1:1).

Data pertaining to interaction effect of time and soil media for planting cuttings recorded maximum (8.58) number of shoots in Vermiculite + sand + FYM (1:1:1) during 15th-21st of July which was followed by (8.08) in Vermiculite + sand (1:1) while, minimum (5.99) in sand + FYM (1:1). Similarly, data obtained during 15th-21st of June recorded maximum (8.37) number of shoots in Vermiculite + sand + FYM (1:1:1) which was followed by (7.99) in Vermiculite + sand (1:1) while, minimum (5.88) in sand + FYM (1:1).

4.2.3.6 Number of Shoots per Plant at 180 DAP

The data pertaining to influence of time and soil media for guava propagation through cuttings on number of shoots per plant after 180 days presented in Table 18 showed significant differences among all the treatments.

4.2.3.6.1 Influence of Time and Soil Media for Planting Cuttings on Number of Shoots per Plant in Guava cv. L-49

Data presented in Table 18 showed maximum (9.82) average number of shoots in treatment combination vermiculite + sand + FYM (1:1) which was followed by (9.57) in Vermiculite + sand (1:1) while, minimum (7.07) in sand + FYM (1:1). At 180 days, the maximum (9.34) average number of shoot was observed in cuttings planted during 15th-21st of August which was followed by (9.16) in cuttings planted during 15th-21st of September while, minimum (8.08) during 15th-21st of June.

4.2.3.6.2 Interaction Effect of Time and Media for Planting Cuttings on Number of Shoots per Plant in Guava cv. L-49

The interaction effect of time and media for guava propagation through cuttings during 15th-21st of August recorded maximum (10.42) number of shoots in Vermiculite + sand + FYM (1:1:1) which was followed by (9.59) in Vermiculite + sand (1:1) while, minimum (7.30) in sand + FYM (1:1). Similarly, during 15th-21st of September maximum (10.19) number of shoots was recorded in Vermiculite + sand + FYM (1:1:1) which was followed by (10.00) in Vermiculite + sand (1:1) while, minimum (7.21) in sand + FYM (1:1).

Data pertaining to interaction effect of soil media and time of propagation through cuttings during 15th-21st of July recorded maximum (10.00) number of shoots in Vermiculite + sand + FYM (1:1:1) which was followed by (9.59) in Vermiculite + sand (1:1) while, minimum (7.11) in sand + FYM (1:1). Similarly, during 15th-21st of June maximum (8.67) number of shoots was recorded in Vermiculite + sand + FYM (1:1:1) which was followed by (8.58) in Vermiculite + sand (1:1:1) while, minimum (6.66) in sand + FYM (1:1).

4.2.4.1 Plant Height (cm)

The data pertaining to the effect of time and soil media for guava propagation through cuttings on plant height at 30, 60, 90, 120, 150 and 180 days after planting are presented in Table 19.

4.2.4.1 Plant Height at 30 DAP

The data obtained after 30 days on the influence of time and soil media on plant height of guava cv. L-49 showed non-significant differences presented in Table 19.

4.2.4.2 Plant Height at 60 DAP

The data presented in Table 19 showed that time and soil media for planting cuttings had after 60 days showed significant effect on plant height of guava.

4.2.4.2.1 Influence of Time and Soil Media for Planting Cuttings on Plant Height (cm) in Guava cv. L-49.

The data presented in Table 19 showed maximum (16.56 cm) average plant height in treatment combination vermiculite + sand + FYM (1:1:1) which was followed by (15.55 cm) in Vermiculite + sand (1:1) while, minimum (12.51 cm) in sand + FYM (1:1). After 60 days maximum (15.55 cm) average plant height was observed during 15th-21st of August which was followed (15.29 cm) during 15th-21st of September while, minimum (13.29 cm) during 15th-21st of June.

4.2.4.2.2 Interaction Effect of Time and Soil Media for Planting Cuttings Plant Height (cm) in Guava cv. L-49

The data presented in Table 19 on interaction effect of time and soil media for guava propagation through cuttings showed maximum (17.57 cm) plant height during 15th-21st of August in Vermiculite + sand + FYM (1:1:1) which was followed by (16.59 cm) in Vermiculite + sand (1:1) while, minimum (13.80 cm) in sand + FYM (1:1). Similarly, during 15th-21st of September maximum (17.57 cm) plant height was observed in Vermiculite + sand + FYM (1:1:1) which was followed by (16.36 cm) in Vermiculite + sand (1:1) while, minimum (13.20 cm) in sand + FYM (1:1).

Data pertaining to interaction effect of time and soil media for guava propagation through cuttings during 15th-21st of July recorded maximum (15.96 cm) plant height in Vermiculite + sand + FYM (1:1:1) which was followed by (14.89 cm) in Vermiculite + sand (1:1) while, minimum (11.87 cm) in sand + FYM (1:1). Similarly, during 15th-21st of June maximum (15.36 cm) plant height was recorded in Vermiculite + sand + FYM (1:1:1) which was followed by (14.36 cm) in Vermiculite + sand (1:1) while, minimum (11.18 cm) in sand + FYM (1:1).

4.2.4.3 Plant Height at 90 DAP

The data pertaining to influence of time and soil media for guava propagation through cuttings on plant height after 90 days presented in Table 19 showed significant differences among all the treatments.

4.2.4.3.1 Influence of Time and Soil Media for Planting Cuttings on Plant Height (cm) in Guava cv. L-49

The data presented in Table 19 showed maximum (18.56 cm) average plant height in treatment combination vermiculite + sand + FYM (1:1:1) which was followed by (17.23 cm) in Vermiculite + sand (1:1) while, minimum (13.54 cm) in sand + FYM (1:1). At 90 days, the maximum (18.52 cm) average plant height was observed during 15th-21st of August which was followed by (17.04 cm) during 15th-21st of September while, minimum (14.12 cm) during 15th-21st of June.

4.2.4.3.2 Interaction Effect of Time and Soil Media for Planting Cuttings on Plant Height (cm) in Guava cv. L-49

The data presented in Table 19 on interaction effect of time and soil media for guava propagation through cuttings recorded maximum (21.17 cm) plant height during 15th-21st of August in treatment combination Vermiculite + sand + FYM (1:1:1) which was followed by (19.40 cm) in Vermiculite + sand (1:1) while, minimum (15.43 cm) in sand + FYM (1:1). Similarly, data obtained during 15th-21st of September showed maximum (19.30 cm) plant height in Vermiculite + sand + FYM (1:1:1) which was followed by (18.10 cm) in Vermiculite + sand (1:1) while, minimum (14.47 cm) in sand + FYM (1:1).

Table 19: Influence of Soil Media, Time of Planting Cuttings and their Interaction Effect on Plant Height (cm) of Guava cv. L-49

Time of Propagation	*Cuttings Soil Media*														
	June 15th-21st	*July 15th-21st*	*Aug 15th-21st*	*Sep 15th-21st*	*Mean*	*June 15th-21st*	*July 15th-21st*	*Aug 15th-21st*	*Sep 15th-21st*	*Mean*	*June 15th-21st*	*July 15th-21st*	*Aug 15th-21st*	*Sep 15th-21st*	*Mean*
	After 30 Days					*After 60 Days*					*After 90 Days*				
Vermiculite + Sand	13.10	13.43	14.67	14.04	13.81	14.36	14.89	16.59	16.36	15.55	15.05	16.40	19.40	18.10	17.23
Perlite + Sand	10.07	10.43	12.07	12.03	11.15	12.17	12.30	14.49	14.26	13.30	13.10	13.40	17.37	16.21	15.02
Sand + FYM	9.00	9.33	10.20	10.00	9.63	11.18	11.87	13.80	13.20	12.51	12.02	12.24	15.43	14.47	13.54
Vermiculite + Sand + FYM	13.80	14.70	15.77	15.13	14.85	15.36	15.96	17.57	17.57	16.56	16.40	17.37	21.17	19.30	18.56
Perlite + Sand+ FYM	12.40	12.77	14.80	13.47	13.36	13.39	13.86	15.33	15.28	14.47	14.04	14.10	19.23	17.12	16.12
Mean	11.67	12.13	13.50	12.93		13.29	13.77	15.55	15.29		14.12	14.70	18.52	17.04	

Factors	*S.E. m (±)*	*C.D (0.05)*	*S.E. m (±)*	*C.D (0.05)*	*S.E. m (±)*	*C.D (0.05)*
Soil media		N.S	0.04	0.11	0.19	0.54
Time of planting cuttings		N.S	0.03	0.11	0.17	0.48
Media x Time		**N.S**	**0.08**	**0.22**	**0.37**	**1.07**

Contd...

Table 19–*Contd...*

Time of Propagation	*Cuttings Soil Media*														
	June 15th-21st	*July 15th-21st*	*Aug 15th-21st*	*Sep 15th-21st*	*Mean*	*June 15th-21st*	*July 15th-21st*	*Aug 15th-21st*	*Sep 15th-21st*	*Mean*	*June 15th-21st*	*July 15th-21st*	*Aug 15th-21st*	*Sep 15th-21st*	*Mean*
	After 120 Days					*After 150 Days*					*After 180 Days*				
Vermiculite +Sand	15.61	18.37	21.38	19.27	18.66	17.48	19.17	23.59	20.17	20.10	18.05	20.13	24.54	21.15	20.97
Perlite + Sand	13.70	14.27	19.20	16.35	15.88	16.19	17.37	20.26	19.15	18.24	17.20	18.16	22.16	20.24	19.44
Sand + FYM	13.33	12.48	17.59	15.17	14.64	15.37	16.15	19.16	18.19	17.22	16.59	17.20	21.13	19.21	18.56
Vermiculite + Sand + FYM	16.58	19.49	23.39	21.38	20.21	18.38	20.17	25.38	22.41	21.59	20.53	21.22	26.49	23.17	22.85
Perlite + Sand+ FYM	14.17	16.15	20.19	18.20	17.18	17.97	18.07	22.18	20.05	19.57	18.20	19.15	23.19	21.17	20.43
Mean	15.88	16.15	20.35	18.07		17.08	18.18	22.11	19.99		18.11	19.17	23.52	20.99	

Factors	*S.E. m (±)*	*C.D (0.05)*	*S.E. m (±)*	*C.D (0.05)*	*S.E. m (±)*	*C.D (0.05)*
Soil media	0.01	0.01	0.06	0.18	0.07	0.20
Time of planting cuttings	0.01	0.01	0.06	0.16	0.06	0.17
Media x Time	**0.01**	**0.03**	**0.13**	**0.37**	**0.14**	**0.39**

Data pertaining to interaction effect of soil media and time for guava propagation through cuttings during 15th-21st of July recorded maximum (17.37 cm) plant height in Vermiculite + sand + FYM (1:1:1) which was at par with (16.40 cm) in Vermiculite + sand (1:1) while, minimum (12.24 cm) in sand + FYM (1:1). Similarly, during 15th-21st of June maximum (16.40 cm) plant height was recorded in Vermiculite + sand + FYM (1:1:1) which was at par with (15.05 cm) in Vermiculite + sand (1:1) while, minimum (12.02 cm) in sand + FYM (1:1).

4.2.2.4 Plant Height at 120 DAP

The data pertaining to influence of time and soil media for guava propagation through cuttings on plant height after 120 days presented in Table 19 showed significant differences among all the treatments.

4.2.2.4.1 Influence of Time and Soil Media for Planting Cuttings on Plant Height (cm) in Guava cv. L-49

The data presented in Table 19 showed maximum (20.21 cm) average plant height in treatment combination vermiculite + sand + FYM (1:1:1) which was followed by (18.66 cm) in Vermiculite + sand (1:1) while, minimum (14.64 cm) in sand + FYM (1:1). At 120 Days, the maximum (20.35 cm) average plant height was observed during 15th-21st of August which was followed by (18.07 cm) during 15th-21st of September while, minimum (15.88 cm) during 15th-21st of June.

4.2.2.4.1 Interaction Effect of Time and Soil Media for Planting Cuttings on Plant Height (cm) in Guava cv. L-49

The data presented in Table 19 on interaction effect of time and soil media for guava propagation through cuttings recorded maximum (23.39 cm) plant height in treatment combination Vermiculite + sand + FYM (1:1:1) during 15th-21st of August which was followed by (21.38 cm) in Vermiculite + sand (1:1) while, minimum (17.59 cm) in sand + FYM (1:1). Similarly, during 15th-21st of September maximum (21.38 cm) plant height was recorded in Vermiculite + sand + FYM (1:1:1) which was followed by (19.27 cm) in Vermiculite + sand (1:1) while, minimum (15.17 cm) in sand + FYM (1:1).

Data pertaining to interaction effect of time and soil media for guava propagation through cuttings during 15th-21st of July recorded maximum (19.49 cm) plant height in treatment combination Vermiculite + sand + FYM (1:1:1) which was followed by (18.37 cm) in Vermiculite + sand (1:1) while, minimum (12.48 cm) in sand + FYM (1:1). Similarly, during 15th-21st of June maximum (16.58 cm) plant height was recorded in Vermiculite + sand + FYM (1:1:1) which was followed by (15.61 cm) in Vermiculite + sand (1:1) while, minimum (13.33 cm) in sand + FYM (1:1).

4.2.4.5 Plant Height at 150 DAP

The data pertaining to influence of time and soil media for guava propagation through cuttings on plant height after 150 days presented in Table 19 showed significant differences among all the treatments.

4.2.4.5.1 Influence of Time and Soil Media for Planting Cuttings on Plant Height (cm) in Guava cv. L-49

The data presented in Table 19 showed maximum (21.59 cm) average plant height in vermiculite + sand + FYM (1:1:1) which was followed by (20.10 cm) in Vermiculite + sand (1:1) while, minimum (17.22 cm) in sand + FYM (1:1). At 150 days maximum (22.11 cm) average plant height was observed during 15th-21st of August which was followed by (19.99 cm) during 15th-21st of September while, minimum (17.08 cm) during 15th-21st of June.

4.2.4.5.2 Interaction Effect of Time and Soil Media for Planting Cuttings on Plant Height (cm) in Guava cv. L-49

The data presented in Table 19 on interaction effect of time and soil media for guava propagation through cuttings during 15th-21st of August recorded maximum (25.38 cm) plant height in treatment combination Vermiculite + sand + FYM (1:1:1) which was followed by (23.59 cm) in Vermiculite + sand (1:1) while, minimum (19.16 cm) in sand + FYM (1:1). Similarly, during 15th-21st of September maximum (22.41 cm) plant height was recorded in Vermiculite + sand + FYM (1:1:1) which was followed by (20.17 cm) in Vermiculite + sand (1:1) while, minimum (18.19 cm) in sand + FYM (1:1).

Data pertaining to interaction effect of time and soil media for guava propagation through cuttings during 15th-21st of July showed maximum (16.15 cm) plant height in Vermiculite + sand + FYM (1:1:1) which was followed by (19.17 cm) in Vermiculite + sand (1:1) while, minimum (16.15 cm) in sand + FYM (1:1). Similarly, during 15th-21st of June maximum (18.38 cm) plant height was recorded in Vermiculite + sand + FYM (1:1:1) which was followed by (17.48 cm) in Vermiculite + sand (1:1) while, minimum (15.37 cm) in sand + FYM (1:1).

4.2.4.6 Plant Height at 180 DAP

The data pertaining to influence of time and soil media for guava propagation through cuttings on plant height after 180 days presented in Table 19 showed significant differences among all the treatments.

4.2.4.6.1 Influence of Time and Soil Media for Planting Cuttings on Plant Height (cm) in Guava cv. L-49

The data presented in Table 19 showed maximum (22.85 cm) average plant height in treatment combination vermiculite + sand + FYM (1:1:1) which was followed by (20.97 cm) in Vermiculite + sand (1:1) while, minimum (18.56 cm) plant height was found in sand + FYM (1:1). At 180 days, the maximum (23.52 cm) average plant height was observed during 15th-21st of August which was followed by (20.99 cm) during 15th-21st of September while, minimum (18.11 cm) during 15th-21st of June.

4.2.4.6.2 Interaction Effect of Time and Soil Media for Planting Cuttings on Plant Height (cm) in Guava cv. L-49

The data presented in Table 19 on interaction effect of time and soil media for guava propagation through cuttings during 15th-21st of August recorded maximum (26.49 cm) plant height in Vermiculite + sand + FYM (1:1:1) which was followed by

(24.54 cm) in Vermiculite + sand (1:1) while, minimum (21.13 cm) in sand + FYM (1:1). Similarly, during 15th-21st of September maximum (23.17 cm) plant height was recorded in Vermiculite + sand + FYM (1:1:1) which was followed by (21.15 cm) in Vermiculite + sand (1:1) while, minimum (19.21 cm) in sand + FYM (1:1).

Data pertaining to interaction effect of soil media and time for guava propagation through cuttings during 15th-21st of July recorded maximum (21.22 cm) plant height in treatment combination Vermiculite + sand + FYM (1:1:1) which was followed by (20.13 cm) in Vermiculite + sand (1:1) while, minimum (17.20 cm) in sand + FYM (1:1). Similarly, during 15th-21st of June maximum (20.53 cm) plant height was recorded in Vermiculite + sand + FYM (1:1:1) which was followed by (18.05 cm) in Vermiculite + sand while, minimum (16.59 cm) in sand + FYM (1:1).

4.1.5 Stem Thickness (cm)

The data pertaining to the effect of time and soil media for propagation of guava through cuttings on stem thickness at 30, 60, 90, 120, 150, 180 days are presented in Table 20.

4.2.5.1 Stem Thickness at 30 DAP

The data obtained after 30 days on the influence of time and soil media on stem thickness (cm) of guava cv. L-49 showed non-significant differences presented in Table 20.

4.2.5.2 Stem thickness at 60 DAP (cm)

The data pertaining to influence of time and soil media for guava propagation through cuttings on stem thickness at 60 days presented in Table 20 showed significant differences among all the treatments.

4.2.5.2.1 Influence of Time and Soil Media for Planting Cuttings on Stem Thickness (cm) in Guava cv. L-49

The data presented in Table 20 showed maximum (0.99 cm) average stem thickness in treatment combination vermiculite + sand + FYM (1:1:1) which was at par with (0.98 cm) in Vermiculite + sand (1:1) while, minimum (0.91 cm) in sand + FYM (1:1). At 60 days, maximum (0.98 cm) average stem thickness was observed during 15th-21st of August which was at par with (0.97 cm) during 15th-21st of September while, minimum (0.91 cm) during 15th-21st of June.

4.2.5.2.2 Interaction Effect of Time and Soil Media for Planting Cuttings on Stem Thickness (cm) in Guava cv. L-49

The data presented in Table 20 on interaction effect of time and soil media for guava propagation through cuttings recorded maximum (1.03 cm) stem thickness in treatment combination Vermiculite + sand + FYM (1:1:1) and Vermiculite + sand (1:1) which was followed by (0.99cm) in perlite + sand + FYM (1:1:1) while, minimum (0.96 cm) in sand + FYM (1:1) during 15th-21st of August. Similarly, during 15th-21st of September maximum (1.00 cm) stem thickness was observed in Vermiculite + sand + FYM (1:1:1) which was followed by (0.98 cm) in Vermiculite + sand (1:1) while, minimum (0.95 cm) in sand + FYM (1:1) and perlite + sand (1:1).

Table 20: Influence of Soil Media, Time of Planting Cuttings and their Interaction Effect on Stem Thickness (cm) of Guava cv. L-49

Time of Propagation	*Cuttings Soil Media*														
	June 15th-21st	*July 15th-21st*	*Aug 15th-21st*	*Sep 15th-21st*	*Mean*	*June 15th-21st*	*July 15th-21st*	*Aug 15th-21st*	*Sep 15th-21st*	*Mean*	*June 15th-21st*	*July 15th-21st*	*Aug 15th-21st*	*Sep 15th-21st*	*Mean*
	After 30 Days					*After 60 Days*					*After 90 Days*				
Vermiculite + Sand	0.71	0.79	0.86	0.84	0.80	0.94	0.96	1.03	0.98	0.98	0.96	0.97	1.04	1.01	1.00
Perlite + Sand	0.69	0.77	0.83	0.81	0.77	0.89	0.92	0.98	0.95	0.93	0.92	0.95	0.92	0.96	0.95
Sand + FYM	0.65	0.74	0.80	0.77	0.74	0.86	0.86	0.96	0.95	0.91	0.88	0.91	0.98	0.94	0.93
Vermiculite + Sand + FYM	0.73	0.81	0.89	0.86	0.82	0.95	0.98	1.03	1.00	0.99	0.98	0.99	1.07	1.04	1.02
Perlite + Sand+ FYM	0.68	0.76	0.83	0.80	0.77	0.92	0.95	0.99	0.97	0.96	0.93	0.96	0.99	0.99	0.97
Mean	0.69	0.77	0.84	0.81		0.91	0.93	0.98	0.97		0.93	0.96	1.00	1.00	

Factors	*S.E. m (±)*	*C.D (0.05)*	*S.E. m (±)*	*C.D (0.05)*	*S.E. m (±)*	*C.D (0.05)*
Soil media		N.S	0.01	0.01	0.01	0.01
Time of planting cuttings		N.S	0.01	0.01	0.01	0.01
Media x Time		**N.S**	**0.01**	**0.02**	**0.01**	**0.02**

Contd...

Table 20–*Contd...*

Time of Propagation	*Cuttings Soil Media*														
	June 15th-21st	*July 15th-21st*	*Aug 15th-21st*	*Sep 15th-21st*	*Mean*	*June 15th-21st*	*July 15th-21st*	*Aug 15th-21st*	*Sep 15th-21st*	*Mean*	*June 15th-21st*	*July 15th-21st*	*Aug 15th-21st*	*Sep 15th-21st*	*Mean*
	After 120 Days					*After 150 Days*					*After 180 Days*				
After 120 days After 150 days After 180 days															
Vermiculite +Sand	0.98	1.00	1.06	1.04	1.02	1.00	1.02	1.07	1.06	1.04	1.03	1.04	1.09	1.07	1.06
Perlite + Sand	0.95	0.98	1.01	0.99	0.98	0.96	1.01	1.04	1.02	1.01	0.99	1.07	1.06	1.03	1.04
Sand + FYM	0.90	0.96	1.00	0.97	0.96	0.94	1.00	1.04	1.01	1.00	0.96	1.01	1.06	1.04	1.02
Vermiculite + Sand + FYM	1.00	1.00	1.09	1.05	1.04	1.04	1.04	1.11	1.07	1.06	1.05	1.06	1.14	1.10	1.08
Perlite + Sand+ FYM	0.94	0.98	1.01	0.99	0.98	0.97	1.03	1.05	1.02	1.02	1.03	1.07	1.07	1.06	1.05
Mean	0.95	0.98	1.01	1.01		0.98	1.02	1.06	1.04		1.02	1.05	1.08	1.06	

Factors	*S.E. m (±)*	*C.D (0.05)*	*S.E. m (±)*	*C.D (0.05)*	*S.E. m (±)*	*C.D (0.05)*
Soil media	0.01	0.01	0.01	0.01	0.01	0.01
Time of planting cuttings	0.01	0.01	0.01	0.01	0.01	0.01
Media x Time	**0.01**	**0.02**	**0.01**	**0.01**	**0.01**	**0.03**

Data pertaining to interaction effect of time and soil media for guava propagation through cuttings during 15th-21st of July recorded maximum (0.98 cm) stem thickness in Vermiculite + sand + FYM (1:1:1) which was at par with (0.96 cm) in Vermiculite + sand (1:1) while, minimum (0.86 cm) sand + FYM (1:1). Similarly, during 15th-21st of June maximum (0.95 cm) stem thickness was observed in Vermiculite + sand + FYM (1:1:1) which at par with (0.94 cm) in Vermiculite + sand (1:1) while, minimum (0.86 cm) in sand + FYM (1:1).

4.2.5.3 Stem Thickness at 90 DAP

The data pertaining to influence of time and soil media for guava propagation through cuttings on stem thickness at 90 days presented in Table 20 showed significant differences among all the treatments.

4.2.5.3.1 Influence of Time and Soil Media for Planting Cuttings on Stem Thickness (cm) in Guava cv. L-49

The data presented in Table 20 showed maximum (1.02 cm) average stem thickness in treatment combination vermiculite + sand + FYM (1:1:1) which was followed by (1.00 cm) in Vermiculite + sand (1:1) while, minimum (0.93 cm) in sand + FYM (1:1). At 90 days maximum (1.00 cm) average stem thickness was observed in cuttings planted during 15th-21st of August and 15th-21st of September while, minimum (0.93 cm) during 15th-21st of June.

4.2.5.3.2 Interaction Effect of Time and Soil Media for Planting Cuttings on Stem Thickness (cm) in Guava cv. L-49

The data presented in Table 20 on interaction effect of time and soil media for guava propagation through cuttings during 15th-21st of August recorded maximum (1.07 cm) stem thickness in treatment combination Vermiculite + sand + FYM (1:1:1) which was followed by (1.04 cm) in Vermiculite + sand (1:1) while, minimum (0.98 cm) in sand + FYM (1:1). Similarly, during 15th-21st of September maximum (1.04 cm) stem thickness was recorded in Vermiculite + sand + FYM (1:1:1) which was followed by (1.01 cm) in Vermiculite + sand (1:1) while, minimum (0.94 cm) in sand + FYM (1:1) and perlite + sand (1:1).

Data pertaining to interaction effect of time and soil media for guava propagation through cuttings during 15th-21st of July recorded maximum (0.99 cm) stem thickness in treatment combination Vermiculite + sand + FYM (1:1:1) which was at par with (0.97 cm) in Vermiculite + sand (1:1) while, minimum (0.91 cm) in sand + FYM (1:1). Similarly, during 15th-21st of June maximum (0.98 cm) stem thickness was recorded in Vermiculite + sand + FYM (1:1:1) which was at par with (0.96 cm) in Vermiculite + sand (1:1) while, minimum (0.88 cm) in sand + FYM (1:1).

4.2.5.4 Stem Thickness at 120 DAP

The data pertaining to influence of time and soil media for guava propagation through cuttings on stem thickness at 120 days presented in Table 20 showed significant differences among all the treatments.

4.2.5.4.1 Influence of Time and Soil Media for Planting Cuttings on Stem Thickness (cm) in Guava cv. L-49

The data presented in Table 20 showed maximum (1.04 cm) average stem thickness in treatment combination vermiculite + sand + FYM (1:1:1) which was followed by (1.02 cm) in Vermiculite + sand (1:1) while, minimum (0.96 cm) in sand + FYM (1:1). At 120 days, maximum (1.01 cm) average stem thickness was observed during 15th-21st of August and 15th-21st of September while, minimum (0.95 cm) during 15th-21st of June.

4.2.5.4.2 Interaction Effect of Time and Soil Media for Planting Cuttings on Stem Thickness (cm) in Guava cv. L-49

The data presented in Table 20 on interaction effect of time and soil media for guava propagation through cuttings during 15th-21st of August showed maximum (1.09 cm) stem thickness in Vermiculite + sand + FYM (1:1:1) which was at par with (1.06 cm) in Vermiculite + sand (1:1) while, minimum (1.00 cm) in sand + FYM (1:1). Similarly, during 15th-21st of September maximum (1.05 cm) stem thickness was recorded in Vermiculite + sand + FYM (1:1:1) which was at par with (1.04 cm) in Vermiculite + sand (1:1) while, minimum (0.97 cm) in sand + FYM (1:1) and perlite + sand (1:1).

Data pertaining to interaction effect of time and soil media for guava propagation through cuttings during 15th-21st of July recorded maximum (1.00 cm) stem thickness in treatment combination Vermiculite + sand + FYM (1:1:1) and Vermiculite + sand (1:1) which was followed by (0.98 cm) in perlite + sand + FYM (1:1:1) while, minimum (0.96 cm) in sand + FYM (1:1). Similarly, during 15th-21st of June maximum (1.00 cm) stem thickness was observed in Vermiculite + sand + FYM (1:1:1) which was followed by (0.98 cm) in Vermiculite + sand (1:1) while, minimum (0.90 cm) in sand + FYM (1:1).

4.2.5.5 Stem Thickness at 150 DAP

The data pertaining to influence of time and soil media for guava propagation through cuttings on stem thickness at 150 days presented in Table 20 showed significant differences among all the treatments.

4.2.5.5.1 Influence of Time and Soil Media for Planting Cuttings on Stem Thickness (cm) in Guava cv. L-49

The data presented in Table 20 showed maximum (1.06 cm) average stem thickness in treatment combination vermiculite + sand + FYM (1:1:1) which was followed by (1.04 cm) in Vermiculite + sand (1:1) while, minimum (1.01 cm) in sand + FYM (1:1). At 150 days, maximum (1.06 cm) average stem thickness was observed during 15th-21st of August which was followed by (1.04 cm) during 15th-21st of September while, minimum (0.98 cm) during 15th-21st of June.

4.2.5.5.2 Influence of Time and Soil Media for Planting Cuttings on Stem Thickness (cm) in Guava cv. L-49

The data presented in Table 20 on interaction effect of time and soil media for guava propagation through cuttings during 15th-21st of August recorded maximum

(1.11 cm) stem thickness in Vermiculite + sand + FYM (1:1:1) which was followed by (1.07 cm) in Vermiculite + sand (1:1) while, minimum (1.04 cm) in sand + FYM (1:1). Similarly, during 15th-21st of September maximum (1.07 cm) stem thickness was recorded in Vermiculite + sand + FYM (1:1:1) which was at par with (1.06 cm) in Vermiculite + sand (1:1) while, minimum (1.01 cm) in sand + FYM (1:1) and perlite + sand (1:1).

Data pertaining to interaction effect of time and soil media for guava propagation through cuttings during 15th-21st of July recorded maximum (1.04 cm) stem thickness in treatment combination Vermiculite + sand + FYM (1:1:1) which was followed by (1.02 cm) in Vermiculite + sand (1:1) while, minimum (1.00 cm) in sand + FYM (1:1). Similarly, during 15th-21st of June maximum (1.04 cm) stem thickness was recorded in Vermiculite + sand + FYM (1:1:1) which was followed by (1.00 cm) in Vermiculite + sand (1:1) while, minimum (0.94 cm) in sand + FYM (1:1).

4.2.5.6 Stem Thickness at 180 DAP

The data pertaining to influence of time and soil media for guava propagation through cuttings on stem thickness at 180 days presented in Table 20 showed significant differences among all the treatments.

4.2.5.6.1 Influence of Time and Soil Media for Planting Cuttings on Stem Thickness (cm) in Guava cv. L-49

The data presented in Table 20 showed maximum (1.08 cm) average stem thickness in treatment combination vermiculite + sand + FYM (1:1:1) which was followed by (1.06 cm) in Vermiculite + sand (1:1) while, minimum (1.02 cm) in sand + FYM (1:1). At 180 days, the maximum (1.08 cm) average stem thickness was observed during 15th-21st of August which was followed by (1.06 cm) during 15th-21st of September while, minimum (1.02cm) during 15th-21st of June.

4.2.5.6.2 Interaction Effect of Time and Soil Media for Planting Cuttings on Stem Thickness (cm) in Guava cv. L-49

The data presented in Table 20 on interaction effect of time and soil media for guava propagation through cuttings during 15th-21st of August recorded maximum (1.14 cm) stem thickness in Vermiculite + sand + FYM (1:1:1) which was followed by (1.09 cm) in Vermiculite + sand (1:1) while, minimum (1.06 cm) in sand + FYM (1:1) and perlite + sand (1:1). Similarly, during 15th-21st of September maximum (1.10 cm) stem thickness was recorded in Vermiculite + sand + FYM (1:1:1) which was at par with (1.07 cm) in Vermiculite + sand (1:1) while, minimum (1.04 cm) in sand + FYM (1:1) and perlite + sand (1:1).

Data pertaining to interaction effect of time and soil media for guava propagation through cuttings during 15th-21st of July recorded maximum (1.06 cm) stem thickness in treatment combination Vermiculite + sand + FYM (1:1:1) which at par with (1.04 cm) in Vermiculite + sand (1:1) while, minimum (1.01 cm) in sand + FYM (1:1). Similarly, during15th-21st of June maximum (1.05 cm) stem thickness was recorded in Vermiculite + sand + FYM (1:1:1) which was at par with (1.03 cm) in Vermiculite + sand (1:1) while, minimum (0.96 cm) in sand + FYM (1:1).

4.2.6 Number of Leaves per Plant

The data pertaining to the effect of time and soil media for propagation of guava through cuttings on number of leaves per plant at 30, 60, 90, 120, 150, 180 days are presented in Table 21.

4.2.6.1 Number of Leaves per Plant at 30 DAP

The data obtained after 30 days on the influence of time and soil media on number of leaves per plant of guava cv. L-49 showed non-significant differences presented in Table 21.

4.2.6.2 Number of Leaves per Plant at 60 DAP

The data pertaining to influence of time and soil media for guava propagation through cuttings on number of leaves per plant at 60 days presented in Table 21 showed significant differences among all the treatments.

4.2.6.2.1 Influence of Time and Soil Media for Planting Cuttings on Number of Leaves per Plant in Guava cv. L-49.

The data presented in Table 21 showed that maximum (10.14) average number of leaves was recorded in treatment combination vermiculite + sand + FYM (1:1:1) which was followed by (9.11) in Vermiculite + sand (1:1) while, minimum (6.36) was found in sand + FYM (1:1). At 60 days, the maximum (8.62) average number of leaves was observed during 15th-21st of August which was followed by (8.49) during 15th-21st of September while, minimum (8.06) during 15th-21st of June.

4.2.6.2.2 Interaction Effect of Time and Soil Media for Planting Cuttings on Number of Leaves per Plant in Guava cv. L-49

The data presented in Table 21 on interaction effect of time and soil media for guava propagation through cuttings during 15th-21st of August showed maximum (10.40) number of leaves in treatment combination Vermiculite + sand + FYM (1:1:1) which was followed by (10.21) in Vermiculite + sand (1:1) while, minimum (6.38) in sand + FYM (1:1). Similarly, during 15th-21st of September maximum (10.09) number of leaves was recorded in Vermiculite + sand + FYM (1:1:1) which was followed by (10.04) in Vermiculite + sand (1:1) while, minimum (6.33) was observed in sand + FYM (1:1) and perlite + sand (1:1).

Data pertaining to interaction effect of time and soil media for guava propagation through cuttings during 15th-21st of July recorded maximum (10.09) number of leaves in treatment combination Vermiculite + sand + FYM (1:1:1) which followed by (8.11) in Vermiculite + sand (1:1) while, minimum (6.30) in sand + FYM (1:1). Similarly, during 15th-21st of June maximum (9.98) number of leaves was recorded in Vermiculite + sand + FYM (1:1:1) which followed by (8.07) in Vermiculite + sand (1:1) while, minimum (6.45) was in sand + FYM (1:1).

4.1.6.3 Number of Leaves per Plant at 90 DAP

The data pertaining to influence of time and soil media for guava propagation through cuttings on number of leaves per plant at 90 days presented in Table 21 showed significant differences among all the treatments.

Table 21: Influence of Soil Media, Time of Planting Cuttings and their Interaction Effect on Number of Leaves per Plant of Guava cv. L-49

Time of Propagation	*Cuttings Soil Media*														
	June 15th-21st	*July 15th-21st*	*Aug 15th-21st*	*Sep 15th-21st*	*Mean*	*June 15th-21st*	*July 15th-21st*	*Aug 15th-21st*	*Sep 15th-21st*	*Mean*	*June 15th-21st*	*July 15th-21st*	*Aug 15th-21st*	*Sep 15th-21st*	*Mean*
	After 30 Days					*After 60 Days*					*After 90 Days*				
Vermiculite + Sand	4.53	4.54	4.89	4.73	4.68	8.07	8.11	10.21	10.04	9.11	14.11	14.79	15.56	14.94	14.85
Perlite + Sand	4.50	4.52	4.53	4.53	4.52	7.87	7.90	8.04	7.98	7.95	12.19	12.78	13.20	12.99	12.78
Sand + FYM	4.28	4.47	4.48	4.73	4.49	6.45	6.30	6.38	6.33	6.36	11.29	11.59	11.92	11.69	11.62
Vermiculite + Sand + FYM	4.89	4.90	4.91	4.53	4.82	9.98	10.09	10.40	10.09	10.14	15.00	15.09	15.88	15.59	15.39
Perlite + Sand+ FYM	4.51	4.60	4.54	4.61	4.58	7.95	7.97	8.06	8.00	7.99	13.29	13.54	14.64	13.79	13.82
Mean	4.54	4.62	4.67	4.64		8.06	8.07	8.62	8.49		13.17	13.56	14.24	13.80	

Factors	*S.E. m (±)*	*C.D (0.05)*	*S.E. m (±)*	*C.D (0.05)*	*S.E. m (±)*	*C.D (0.05)*
Soil media		N.S	0.02	0.07	0.01	0.02
Time of planting cuttings		N.S	0.02	0.06	0.01	0.02
Media x Time		**N.S**	**0.05**	**0.14**	**0.01**	**0.04**

Contd...

Table 21–*Contd...*

Time of Propagation	*Cuttings Soil Media*														
	June 15th-21st	*July 15th-21st*	*Aug 15th-21st*	*Sep 15th-21st*	*Mean*	*June 15th-21st*	*July 15th-21st*	*Aug 15th-21st*	*Sep 15th-21st*	*Mean*	*June 15th-21st*	*July 15th-21st*	*Aug 15th-21st*	*Sep 15th-21st*	*Mean*
	After 120 Days					*After 150 Days*					*After 180 Days*				
Vermiculite +Sand	15.00	15.19	16.26	16.19	15.66	20.49	20.88	21.22	21.10	20.92	23.49	23.88	24.19	23.99	23.89
Perlite + Sand	15.00	15.04	15.87	15.58	15.38	18.79	19.38	20.99	20.65	19.95	23.04	23.22	23.78	23.56	23.41
Sand + FYM	14.17	14.60	14.89	14.72	14.60	17.27	17.37	17.77	17.61	17.50	19.82	19.87	20.87	20.64	20.30
Vermiculite + Sand + FYM 24.18		18.23	18.38	18.62	18.47	18.42	21.29	21.60	21.82	21.60	21.67	23.58	23.88	25.08	24.19
Perlite + Sand+ FYM	15.08	15.09	16.09	15.79	15.51	19.21	19.99	21.10	20.77	20.27	19.83	24.93	24.29	24.99	23.71
Mean	15.50	15.60	16.35	16.15		19.41	19.84	20.61	20.39		21.96	23.15	23.64	23.47	

Factors	*S.E. m (±)*	*C.D (0.05)*	*S.E. m (±)*	*C.D (0.05)*	*S.E. m (±)*	*C.D (0.05)*
Soil media	0.01	0.02	0.01	0.01	0.01	0.01
Time of planting cuttings	0.01	0.02	0.01	0.01	0.01	0.01
Media x Time	**0.01**	**0.04**	**0.01**	**0.03**	**0.01**	**0.03**

4.2.6.3.1 Influence of Time and Soil Media for Planting Cuttings on Number of Leaves per Plant in Guava cv. L-49

The data presented in Table 21 on interaction effect of time and soil media for guava propagation through cuttings recorded maximum (15.39) average number of leaves treatment combination vermiculite + sand + FYM (1:1:1) which was followed by (14.85) in Vermiculite + sand (1:1) while, minimum (11.62) in sand + FYM (1:1). At 90 days, the maximum (14.24) average number of leaves was observed during 15th-21st of August which was followed by (13.80) during 15th-21st of September while, minimum (13.17) during 15th-21st of June.

4.2.6.3.2 Interaction Effect of Time and Soil Media for Planting Cuttings on Number of Leaves per Plant in Guava cv. L-49

The data presented in Table 21 on interaction effect of time and soil media for guava propagation through cuttings during 15th-21st of August recorded maximum (15.88) number of leaves in Vermiculite + sand + FYM (1:1:1) which was followed by (15.56) in Vermiculite + sand (1:1:1) while, minimum (11.92) in sand + FYM (1:1). Similarly, during 15th-21st of September maximum (15.59) number of leaves was recorded in Vermiculite + sand + FYM (1:1:1) which was followed by (14.94) in Vermiculite + sand (1:1) while, minimum (11.69) in sand + FYM (1:1) and perlite + sand (1:1).

Data pertaining to interaction effect of time and soil media for guava propagation through cuttings during 15th-21st of July recorded maximum (15.09) number of leaves in treatment combination Vermiculite + sand + FYM (1:1:1) which followed by (14.79) in Vermiculite + sand (1:1) while, minimum (11.59) in sand + FYM (1:1). Similarly, during 15th-21st of June maximum (15.00) number of leaves was recorded in Vermiculite + sand + FYM (1:1:1) which followed by (14.11) in Vermiculite + sand (1:1) while, minimum (11.29) in sand + FYM (1:1).

4.1.6.4 Number of Leaves per Plant at 120 DAP

The data pertaining to influence of time and soil media for guava propagation through cuttings on number of leaves per plant at 120 days presented in Table 21 showed significant differences among all the treatments.

4.2.6.4.1 Influence of Time and Soil Media for Planting Cuttings on Number of Leaves per Plant in Guava cv. L-49

The data presented in Table 21 showed maximum (18.42) average number of leaves in treatment combination vermiculite + sand + FYM (1:1:1) which was followed by (15.66) Vermiculite + sand (1:1) while, minimum (14.60) in sand + FYM (1:1). At 120 days, the maximum average (16.35) number of leaves was observed during 15th-21st of August which followed by (16.15) during 15th-21st of September while, minimum (15.50) during 15th-21st of June.

4.2.6.4.2 Interaction Effect of Time and Soil Media for Planting Cuttings on Number of Leaves per Plant in Guava cv. L-49

The data presented in Table 21 on interaction effect of time and soil media for guava propagation through cuttings during 15th-21st of August showed maximum

(18.62) number of leaves in Vermiculite + sand + FYM (1:1:1) which was followed by (16.26) in Vermiculite + sand (1:1) while, minimum (14.89) in sand + FYM (1:1). Similarly, during 15th-21st of September maximum (18.47) number of leaves was recorded in Vermiculite + sand + FYM (1:1:1) which followed by (16.19) in Vermiculite + sand (1:1) while, minimum (14.72) in sand + FYM (1:1) and perlite + sand (1:1).

Data pertaining to interaction effect of time and soil media for guava propagation through cuttings during 15th-21st of July recorded maximum (18.38) number of leaves in Vermiculite + sand + FYM (1:1:1) which followed by (15.19) in Vermiculite + sand (1:1) while, minimum (14.60) in sand + FYM (1:1). Similarly, during 15th-21st of June maximum (18.23) number of leaves was recorded in Vermiculite + sand + FYM (1:1:1) which followed by (15.00) in Vermiculite + sand (1:1) while, minimum (14.17) in sand + FYM (1:1).

4.1.6.5 Number of Leaves per Plant at 150 DAP

The data pertaining to influence of time and soil media for guava propagation through cuttings on number of leaves per plant at 150 days presented in Table 21 showed significant differences among all the treatments.

4.2.6.5.1 Influence of Time and Soil Media for Planting Cuttings on Number of Leaves per Plant in Guava cv. L-49

The data presented in Table 21 on interaction effect of time and soil media for guava propagation through cuttings recorded maximum (21.67) average number of leaves in treatment combination vermiculite + sand + FYM (1:1:1) which was followed by (20.92) in Vermiculite + sand (1:1) while, minimum (17.20) in sand + FYM (1:1). At 150 days, the maximum (20.58) average number of leaves was observed during 15th-21st of August which followed by (20.34) during 15th-21st of September while, minimum (19.41) number of leaves was recorded during 15th-21st of June.

4.2.6.5.2 Interaction Effect of Time and Soil Media for Planting Cuttings on Number of Leaves per Plant in Guava cv. L-49

The data presented in Table 21 on interaction effect of time and soil media for guava propagation through cuttings during 15th-21st of August recorded maximum (21.82) number of leaves in treatment combination Vermiculite + sand + FYM (1:1:1) which was followed by (21.22) in Vermiculite + sand (1:1) while, minimum (21.22) in sand + FYM (1:1). Similarly, during 15th-21st of September maximum (21.60) number of leaves was recorded in Vermiculite + sand + FYM (1:1:1) which was followed by (21.10) in Vermiculite + sand (1:1) while, minimum (17.61) in sand + FYM (1:1) and perlite + sand (1:1).

Data pertaining to interaction effect of time and soil media for guava propagation through cuttings during 15th-21st of July recorded maximum (21.60) number of leaves in Vermiculite + sand + FYM (1:1:1) which was followed by (20.88) in Vermiculite + sand (1:1) while, minimum (17.37) number of leaves in sand + FYM (1:1). Similarly, during 15th-21st of June maximum (21.29) number of leaves was recorded in Vermiculite + sand + FYM (1:1:1) which followed by (20.49) in Vermiculite + sand (1:1) while, minimum (17.27) in sand + FYM (1:1).

4.1.6.6 Number of Leaves per Plant at 180 DAP

The data pertaining to influence of time and soil media for guava propagation through cuttings on number of leaves per plant at 180 days presented in Table 21 showed significant differences among all the treatments.

4.2.6.6.1 Influence of Time and Soil Media for Planting Cuttings on Number of Leaves per Plant in Guava cv. L-49

The data presented in Table 21 showed maximum (23.98) average number of leaves in treatment combination vermiculite + sand + FYM (1:1:1) which was followed by (23.89) in Vermiculite + sand (1:1) while, minimum (20.30) in sand + FYM (1:1). At 180 days, maximum (23.64) average number of leaves was observed during 15th-21st of August which was followed by (23.47) during 15th-21st of September while, minimum (21.96) was recorded during 15th-21st of June.

4.2.6.6.2 Interaction Effect of Time and Soil Media for Planting Cuttings on Number of Leaves per Plant in Guava cv. L-49

The data presented in Table 21 on interaction effect of time and soil media for guava propagation through cuttings during 15th-21st of August recorded maximum (24.29) number of leaves in Vermiculite + sand + FYM (1:1:1) which was followed by (24.19) in Vermiculite + sand (1:1) while, minimum (20.87) in sand + FYM (1:1). Similarly, during 15th-21st of September maximum (24.19) number of leaves was observed in Vermiculite + sand + FYM (1:1:1) which was followed by (23.99) in Vermiculite + sand (1:1) while, minimum (20.64) in sand + FYM (1:1).

Data pertaining to interaction effect of time and soil media for guava propagation through cuttings during 15th-21st of July recorded maximum (23.88) number of leaves in Vermiculite + sand + FYM (1:1:1) and Vermiculite + sand (1:1) while, minimum (19.87) in sand + FYM (1:1). Similarly, during 15th-21st of June maximum (23.58) number of leaves was observed in Vermiculite + sand + FYM (1:1:1) which was followed by (23.49) in Vermiculite + sand (1:1) while, minimum (19.82) in sand + FYM (1:1).

4.2.7 Average Leaf Area (cm^2)

The data pertaining to influence of time and soil media for guava propagation through cuttings on average leaf area at 90 days presented in Table 22 showed significant differences among all the treatments.

4.2.7.1 Influence of Time and Soil Media for Planting Cuttings on Average Leaf Area (cm^2) in Guava cv. L-49

The data presented in Table 22 showed maximum (20.20 cm^2) average leaf area in treatment combination vermiculite + sand + FYM (1:1:1), which was followed by (20.18 cm^2) in vermiculite + FYM (1:1), while minimum (20.09 cm^2) in sand + FYM (1:1). At 90 days the maximum (22.15 cm^2) average leaf area was recorded during 15th-21st of August which was followed by (20.16 cm^2) during 15th-21st of September, while minimum (18.79 cm^2) during 15th-21st of June.

Table 22: Influence of Soil Media and Time of Planting Cuttings on Average Leaf Area (cm^2) of Guava Cuttings cv. L-49 after 90 Days

Time of Planting	*Cuttings Soil Media*				
	June 15th-21st	*July 15th-21st*	*Aug 15th-21st*	*Sep 15th-21st*	*Mean*
Vermiculite + Sand	18.84	19.54	22.16	20.18	20.18
Perlite + Sand	18.76	19.46	22.14	20.17	20.13
Sand + FYM	18.70	19.41	22.14	20.13	20.09
Vermiculite + Sand + FYM	18.87	19.60	22.17	20.18	20.20
Perlite + Sand+ FYM	18.80	19.49	22.14	20.17	20.15
Mean	**18.79**	**19.50**	**22.15**	**20.16**	

Factors	*S.E. m (±)*	*C.D (0.05)*
Soil media	0.01	0.01
Time of planting cuttings	0.01	0.02
Media x Time	**0.01**	**0.04**

4.2.7.2 Interaction Effect of Time and Soil Media for Planting Cuttings on Average Leaf Area (cm^2) in Guava cv. L-49

The data presented in Table 22 on interaction effect of time and soil media for guava propagation through cuttings during 15th-21st of August showed maximum (22.17 cm^2) leaf area in treatment combination Vermiculite + sand + FYM (1:1:1) which was at par with (22.16 cm^2) in Vermiculite + sand (1:1) while, minimum (22.14 cm^2) in sand + FYM (1:1) and perlite + sand (1:1). Similarly, during 15th-21st of September maximum (20.18 cm^2) leaf area was recorded in Vermiculite + sand + FYM (1:1:1) and Vermiculite + sand (1:1) while, minimum (20.13 cm^2) in sand + FYM (1:1).

Data pertaining to interaction effect of time and soil media for guava propagation through cuttings during 15th-21st of July recorded maximum (19.60 cm^2) leaf area in Vermiculite + sand + FYM (1:1:1) which was followed by (19.54 cm^2) in Vermiculite + sand (1:1) while, minimum (19.41 cm^2) in sand + FYM (1:1). Similarly, during 15th-21st of June maximum (18.87 cm^2) leaf area was recorded in Vermiculite + sand + FYM (1:1:1) which was at par with (18.84 cm^2) in Vermiculite + sand (1:1) while, minimum (18.70 cm^2) in sand + FYM (1:1).

4.2.8 Average Leaf Fresh Weight (g)

The data pertaining to influence of time and soil media for guava propagation through cuttings on average leaf fresh weight at 90 days presented in Table 23 showed significant differences among all the treatments.

Table 23: Influence of Soil Media, Time of Planting Cuttings and their Interaction Effect on Average Leaf Fresh Weight (g) of Guava Cuttings cv. L-49 after 90 Days

Time of Planting	*Cuttings Soil Media*				
	June 15th-21st	*July 15th-21st*	*Aug 15th-21st*	*Sep 15th-21st*	*Mean*
Vermiculite + Sand	0.34	0.36	0.42	0.38	0.37
Perlite + Sand	0.30	0.35	0.39	0.35	0.35
Sand + FYM	0.28	0.34	0.36	0.34	0.33
Vermiculite + Sand + FYM	0.36	0.39	0.43	0.40	0.39
Perlite + Sand+ FYM	0.31	0.36	0.41	0.36	0.36
Mean	**0.32**	**0.36**	**0.40**	**0.36**	

Factors	*S.E. m (±)*	*C.D (0.05)*
Soil media	0.01	0.01
Time of planting cuttings	0.01	0.01
Media x Time	**0.01**	**0.02**

4.2.8.1 Influence of Time and Soil Media for Planting Cuttings on Average Leaf Fresh Weight (g) in Guava cv. L-49

The data presented in Table 23 showed maximum (0.39 g) average leaf fresh weight in treatment combination vermiculite + sand + FYM (1:1:1), which was followed by (0.37 g) in vermiculite + sand (1:1) while, minimum (0.33 g) in sand + FYM (1:1). Similarly, during 15th-21st of August maximum (0.40 g) average leaf fresh weight was observed which was followed by (0.36 g) during 15th-21st of September and 15th-21st of July while, minimum (0.32 g) 15th-21st of June.

4.2.8.2 Interaction Effect of Time and Soil Media for Planting Cuttings on Average Leaf Fresh Weight (g) in Guava cv. L-49

The data presented in Table 23 on interaction effect of time and soil media for guava propagation through cuttings during 15th-21st of August recorded maximum (0.43 g) leaf fresh weight in treatment combination Vermiculite + sand + FYM (1:1:1) which was followed by (0.42 g) in Vermiculite + sand (1:1) while minimum (0.36 g) in sand + FYM (1:1). Similarly, during 15th-21st of September maximum (0.40 g) leaf fresh weight was recorded in Vermiculite + sand + FYM (1:1:1) which was followed by (0.38 g) in Vermiculite + sand (1:1) and perlite + sand + FYM (1:1:1) while, minimum (0.34 g) in sand + FYM (1:1).

Data pertaining to interaction effect of time and soil media for guava propagation through cuttings during 15th-21st of July recorded maximum (0.39 g) leaf fresh weight in Vermiculite + sand + FYM (1:1:1:) which was followed by (0.36 g) in Vermiculite + sand (1:1) and perlite + sand + FYM (1:1:1) while minimum (0.34 g) in sand + FYM (1:1). Similarly, during 15th-21st of June maximum (0.36 g) leaf fresh weight was recorded in Vermiculite + sand + FYM (1:1:1) which was followed by (0.34 g) in Vermiculite + sand (1:1) while, minimum (0.28 g) sand + FYM (1:1).

4.2.9 Average Leaf Dry Weight (mg)

The data pertaining to influence of time and soil media for guava propagation through cuttings on average leaf dry weight at 90 days presented in Table 24 showed significant differences among all the treatments.

Table 24: Influence of Soil Media, Time of Planting Cuttings and their Interaction Effect on Average Leaf Dry Weight (mg) of Guava Cuttings cv. L-49 after 90 Days

Time of Planting	*Cuttings Soil Media*				
	June 15th-21st	*July 15th-21st*	*Aug 15th-21st*	*Sep 15th-21st*	*Mean*
Vermiculite + Sand	21.95	22.94	26.90	23.78	23.87
Perlite + Sand	20.34	21.73	24.35	22.87	22.32
Sand + FYM	19.93	20.64	24.17	22.68	21.85
Vermiculite + Sand + FYM	22.60	24.64	27.00	23.98	24.55
Perlite + Sand+ FYM	21.70	22.71	26.02	23.78	23.55
Mean	21.30	22.53	25.69	23.41	

Factors	*S.E. m (±)*	*C.D (0.05)*
Soil media	0.14	0.39
Time of planting cuttings	0.12	0.35
Media x Time	**0.27**	**0.78**

4.2.9.1 Influence of Time and Soil Media for Planting Cuttings on Average Leaf Dry Weight (mg) in Guava cv. L-49

The data presented in Table 24 showed maximum (24.55 mg) average leaf dry weight in treatment combination vermiculite + sand + FYM (1:1:1), which was followed by (23.87 mg) in vermiculite + sand (1:1) while, minimum (21.85 mg) in sand + FYM (1:1). At 90 days maximum (25.69 mg) average leaf dry weight was observed during 15th-21st of August which was followed by (23.41 mg) during 15th-21st of September while, minimum (21.30 mg) during 15th-21st of June.

4.2.9.2 Interaction Effect of Time and Soil Media for Planting Cuttings on Average Leaf Dry Weight (mg) in Guava cv. L-49

The data presented in Table 24 on interaction effect of time and soil media for guava propagation through cuttings during 15th-21st of August recorded maximum (27.00 mg) leaf dry weight in treatment combination Vermiculite + sand + FYM (1:1:1) which was at par with (26.90 mg) Vermiculite + sand (1:1) while minimum (24.17 mg) leaf dry weight in sand + FYM (1:1). Similarly, during 15th-21st of September maximum (23.98 mg) leaf dry weight was recorded in Vermiculite + sand + FYM (1:1:1) which was at par with (23.78 mg) in Vermiculite + sand (1:1) and perlite + sand + FYM (1:1:1) while, minimum (22.68 mg) in sand + FYM (1:1).

Data pertaining to interaction effect of time and soil media for guava propagation through cuttings during 15th-21st of July recorded maximum (24.64 mg) leaf dry weight in treatment combination Vermiculite + sand + FYM (1:1:1) which was followed by (22.94 mg) in Vermiculite + sand (1:1) and perlite + sand + FYM (1:1:1) while, minimum (20.64 mg) leaf dry weight in sand + FYM (1:1). Similarly, during 15th-21st of June maximum (22.60 mg) leaf dry weight was observed in Vermiculite + sand + FYM (1:1:1) which was at par with (21.95 mg) in Vermiculite + sand (1:1) while, minimum (19.93 mg) in sand + FYM (1:1).

4.2.10 Average Plant Fresh Weight (g)

The data pertaining to influence of time and soil media for guava propagation through cuttings on average plant fresh weight at 90 days presented in Table 25 showed significant differences among all the treatments.

Table 25: Influence of Soil Media, Time of Planting Cuttings and their Interaction Effect on Average Plant Fresh Weight (g) of Guava Cuttings cv. L-49 after 90 Days

Time of Planting	*Cuttings Soil Media*				
	June 15th-21st	*July 15th-21st*	*Aug 15th-21st*	*Sep 15th-21st*	*Mean*
Vermiculite + Sand	3.92	4.56	4.68	4.62	4.45
Perlite + Sand	3.33	4.13	4.48	4.17	4.02
Sand + FYM	3.29	4.04	4.39	4.10	3.95
Vermiculite + Sand + FYM	4.01	4.64	4.78	4.72	4.54
Perlite + Sand+ FYM	3.47	4.16	4.66	4.23	4.13
Mean	**3.60**	**4.31**	**4.60**	**4.37**	

Factors	*S.E. m (±)*	*C.D (0.05)*
Soil media	0.01 0.01	
Time of planting cuttings	0.01 0.01	
Media x Time	**0.01 0.02**	

4.2.10.1 Influence of Time and Soil Media for Planting Cuttings Effect on Average Plant Fresh Weight (g) in Guava cv. L-49

The data presented in Table 25 showed maximum (4.54 g) average plant fresh weight in treatment combination vermiculite + sand + FYM (1:1:1), which was followed by (4.45 g) in vermiculite + sand (1:1) while, minimum (3.95 g) in sand + FYM (1:1). At 90 days, maximum (4.60 g) average plant fresh weight was observed during 15th-21st of August which was followed by (4.37 g) during 15th-21st of September while, minimum (3.60 g) during 15th-21st of June.

4.2.10.2 Interaction Effect of Time and Soil Media for Planting Cuttings Effect on Average Plant Fresh Weight (g) in Guava cv. L-49

The data presented in Table 25 on interaction effect of time and soil media for guava propagation through cuttings during 15th-21st of August recorded maximum (4.78 g) plant fresh weight in treatment combination Vermiculite + sand + FYM (1:1:1) which was followed by (4.68 g) in Vermiculite + sand (1:1:1) while minimum

(4.39 g) in sand + FYM (1:1). Similarly, during 15th-21st of September maximum (4.72 g) plant fresh weight was recorded in Vermiculite + sand + FYM (1:1:1) which was followed by (4.62 g) in Vermiculite + sand (1:1) while, minimum (4.10 g) in sand + FYM (1:1).

Data pertaining to interaction effect of time and soil media for guava propagation through cuttings during 15th-21st of July recorded maximum (4.64 g) plant fresh weight in Vermiculite + sand + FYM (1:1:1) which was followed by (4.56 g) in Vermiculite + sand (1:1) while, minimum (4.04 g) in sand + FYM (1:1). Similarly, during 15th-21st of June maximum (4.01 g) plant fresh weight in Vermiculite + sand + FYM (1:1:1) which was followed by (3.92 g) in Vermiculite + sand (1:1) while, minimum (3.29 g) in sand + FYM (1:1).

4.2.11 Average Plant Dry Weight (g)

The data pertaining to influence of time and soil media for guava propagation through cuttings on average plant dry weight at 90 days presented in Table 26 showed significant differences among all the treatments.

Table 26: Influence of Soil Media, Time of Planting Cuttings and their Interaction Effect on Average Plant Dry Weight (g) of Guava Cuttings cv. L-49 after 90 Days

Time of Planting	*Cuttings Soil Media*				
	June 15th-21st	*July 15th-21st*	*Aug 15th-21st*	*Sep 15th-21st*	*Mean*
Vermiculite + Sand	0.16	0.16	0.18	0.17	0.17
Perlite + Sand	0.13	0.14	0.17	0.20	0.16
Sand + FYM	0.11	0.13	0.15	0.14	0.13
Vermiculite + Sand + FYM	0.20	0.21	0.23	0.19	0.20
Perlite + Sand+ FYM	0.14	0.15	0.17	0.16	0.16
Mean	**0.15**	**0.16**	**0.18**	**0.17**	

Factors	*S.E. m (±)*	*C.D (0.05)*
Soil media	0.01	0.01
Time of planting cuttings	0.01	0.01
Media x Time	**0.01**	**0.02**

4.2.11.1 Influence of Time and Soil Media for Planting Cuttings on Average Plant Dry Weight (g) in Guava cv. L-49

The data presented in Table 26 showed maximum (0.20 g) average plant dry weight in treatment combination vermiculite + sand + FYM (1:1:1) which was followed by (0.17 g) in vermiculite + sand (1:1) while, minimum (0.13 g) in sand + FYM (1:1). At 90 days, maximum (0.18 g) average plant dry weight was observed during 15th-21st of August which was followed by (0.17 g) during 15th-21st of September while, minimum (0.15 g) plant dry weight during 15th-21st of June.

4.2.11.2 Interaction Effect of Time and Soil Media for Planting Cuttings on Average Plant Dry Weight (g) in Guava cv. L-49

The data presented in Table 26 on interaction effect of time and soil media for guava propagation through cuttings during 15th-21st of August recorded maximum (0.23 g) plant fresh weight in treatment combination Vermiculite + sand + FYM (1:1:1) which was followed by (0.18 g) in Vermiculite + sand (1:1) while minimum (0.15 g) in sand + FYM (1:1). Similarly, during 15th-21st of September maximum (0.19 g) plant dry weight was recorded in Vermiculite + sand + FYM (1:1:1) which was followed by (0.17 g) in Vermiculite + sand (1:1) while, minimum (0.14 g) in sand + (1:1).

Data pertaining to interaction effect of time and soil media for guava propagation through planting cuttings during 15th-21st of July recorded maximum (0.21 g) plant dry weight in Vermiculite + sand + FYM (1:1:1) which was followed by (0.16 g) in Vermiculite + sand (1:1) while, minimum (0.13 g) in sand + FYM (1:1). Similarly, during 15th-21st of June maximum (0.20 g) plant dry weight was recorded in Vermiculite + sand + FYM (1:1:1) which was followed by (0.16 g) in Vermiculite + sand (1:1) while, minimum (0.11 g) in sand + FYM (1:1).

4.2.12 Number of Primary Roots per Rooted Cutting

The data pertaining to influence of time and soil media for guava propagation through cuttings on number of primary roots per rooted cutting at 90 days presented in Table 27 showed significant differences among all the treatments.

Table 27: Influence of Soil Media, Time of Planting Cuttings and their Interaction Effect on Number of Primary Roots per Pooted Cuttings of Guava cv. L-49 after 90 Days

Time of Planting	*Cuttings Soil Media*				
	June 15th-21st	*July 15th-21st*	*Aug 15th-21st*	*Sep 15th-21st*	*Mean*
Vermiculite + Sand	4.97	5.12	5.25	5.18	5.13
Perlite + Sand	4.78	4.91	5.06	5.02	4.94
Sand + FYM	4.11	4.86	4.98	4.95	4.72
Vermiculite + Sand + FYM	5.06	5.20	5.26	5.20	5.18
Perlite + Sand+ FYM	4.86	4.92	5.12	5.09	5.10
Mean	**4.75**	**5.00**	**5.13**	**5.09**	

Factors	*S.E. m (±)*	*C.D (0.05)*
Soil media	0.04	0.11
Time of planting cuttings	0.03	0.09
Media x Time	**0.07**	**0.21**

4.2.12.1 Influence of Time and Soil Media for Planting Cuttings and on Number of Primary Roots per Rooted Cutting in Guava cv. L-49

The data presented in Table 27 showed maximum (5.18) average number of primary roots in treatment combination vermiculite + sand + FYM (1:1:1), which was at par with (5.13) in vermiculite + sand (1:1) while, minimum (4.72) in sand + FYM (1:1). At 90 days the maximum (5.13) average number of primary roots was observed during 15th-21st of August which was at par with (5.09) during 15th-21st of September and (5.00) during 15th-21st of July while, minimum (4.75) during 15th-21st of June.

4.2.12.2 Interaction Effect of Time and Soil Media for Planting Cuttings and on Number of Primary Roots per Rooted Cutting in Guava cv. L-49

The data presented in Table 27 on interaction effect of time and soil media for guava propagation through cuttings during 15th-21st of August showed maximum (5.26) number of primary roots in treatment combination Vermiculite + sand + FYM (1:1:1) which was at par with (5.25) in Vermiculite + sand (1:1) and (5.12) in perlite + sand + FYM (1:1:1) while, minimum (4.98) in sand + FYM (1:1). Similarly, during 15th-21st of September maximum (5.20) number of primary roots was recorded in Vermiculite + sand + FYM (1:1:1) which was at par with (5.18) in Vermiculite + sand (1:1) and (5.09) in perlite + sand + FYM (1:1:1) while, minimum (4.95) in sand + FYM (1:1).

Data pertaining to interaction effect of time and soil media for guava propagation through cuttings during 15th-21st of July recorded maximum (5.20) number of primary roots in Vermiculite + sand + FYM (1:1:1) which was at par with (5.12) in Vermiculite + sand (1:1) while, minimum (4.86) in sand + FYM (1:1). Similarly, during 15th-21st of June maximum (5.06) number of primary roots was observed in Vermiculite + sand + FYM (1:1:1) which was at par with (4.97) in Vermiculite + sand (1:1), (4.86) in perlite + sand + FYM (1:1:1) and (4.78) in perlite + sand (1:1) while, minimum (4.11) in sand + FYM.

4.2.13 Number of Secondary Roots per Rooted Cuttings

The data pertaining to influence of time and soil media for guava propagation through cuttings on number of secondary roots at 90 days presented in Table 28 showed significant differences among all the treatments.

4.2.13.1 Influence of Time and Soil Media for Planting Cuttings on Number of Secondary Roots per Rooted Cutting in Guava cv. L-49

The data presented in Table 28 showed maximum (9.33) average number of secondary roots in treatment combination vermiculite + sand + FYM (1:1:1), which was followed by (9.29) in vermiculite + sand (1:1) while, minimum (8.88) in sand + FYM (1:1). At 90 days maximum (9.20) average number of secondary roots was observed during 15th-21st of August which was followed by (9.15) during 15th-21st of September while, minimum (9.07) during 15th-21st of June.

Table 28: Influence of Soil Media, Time of Planting Cuttings and their Interaction Effect on Number of Secondary Roots per Rooted Cutting of Guava cv. L-49 after 90 Days

Time of Planting	*Cuttings Soil Media*				
	June 15th-21st	*July 15th-21st*	*Aug 15th-21st*	*Sep 15th-21st*	*Mean*
Vermiculite + Sand	9.24	9.28	9.34	9.31	9.29
Perlite + Sand	8.95	9.03	9.08	9.04	9.02
Sand + FYM	8.81	8.87	8.94	8.90	8.88
Vermiculite + Sand + FYM	9.25	9.28	9.44	9.35	9.33
Perlite + Sand + FYM	9.11	9.15	9.19	9.17	9.15
Mean	**9.07**	**9.12**	**9.20**	**9.15**	

Factors	*S.E. m (±)*	*C.D (0.05)*
Soil media	0.01	0.02
Time of planting cuttings	0.01	0.02
Media x Time	**0.01**	**0.04**

4.2.13.2 Interaction Effect of Time and Soil Media for Planting Cuttings on Number of Secondary Roots per Rooted Cutting in Guava cv. L-49

The data presented in Table 28 on interaction effect of time and soil media for guava propagation through cuttings during 15th-21st of August showed maximum (9.44) number of secondary roots in Vermiculite + sand + FYM (1:1:1) which was followed (9.34) in Vermiculite + sand (1:1) while, minimum (8.94) in sand + FYM (1:1). Similarly, during 15th-21st of September maximum (9.35) number of secondary roots was recorded in Vermiculite + sand + FYM (1:1:1) which was at par with (9.31) in Vermiculite + sand while, minimum (8.90) in sand + FYM (1:1).

Data pertaining to interaction effect of time and soil media for guava propagation through cuttings during 15th-21st of July recorded maximum (9.28) number of secondary roots in Vermiculite + sand + FYM (1:1:1) and Vermiculite + sand (1:1) while, minimum (8.87) was observed in sand + FYM (1:1). Similarly, during 15th-21st of June maximum (9.25) number of secondary roots was observed in Vermiculite + sand + FYM (1:1:1) which was at par with (9.24) in Vermiculite + sand (1:1) while, minimum (8.81) in sand + FYM (1:1).

4.2.14 Length of Primary Root (cm)

The data pertaining to influence of time and soil media for guava propagation through cuttings on length of primary root at 90 days presented in Table 29 showed significant differences among all the treatments.

Table 29: Influence of Soil Media, Time of Planting Cuttings and their Interaction Effect on Length of Primary Root (cm) of Guava Cuttings cv. L-49 after 90 Days

Time of Planting	*Cuttings Soil Media*				
	June 15th-21st	*July 15th-21st*	*Aug 15th-21st*	*Sep 15th-21st*	*Mean*
Vermiculite + Sand	4.00	4.37	4.68	4.60	4.41
Perlite + Sand	3.46	3.63	4.08	3.74	3.73
Sand + FYM	3.23	3.31	3.81	3.60	3.48
Vermiculite + Sand + FYM	4.48	4.60	4.93	4.76	4.69
Perlite + Sand+ FYM	3.93	4.00	4.53	4.47	4.23
Mean	**3.82**	**3.98**	**4.41**	**4.23**	

Factors	*S.E. m (±)*	*C.D (0.05)*
Soil media	0.03	0.08
Time of planting cuttings	0.02	0.07
Media x Time	**0.06**	**0.16**

4.2.14.1 Influence of Time and Soil Media for Planting Cuttings on Length of Primary Root (cm) in Guava cv. L-49

The data presented in Table 29 showed maximum (4.69 cm) average length of primary root in treatment combination vermiculite + sand + FYM (1:1:1), which was followed by (4.41 cm) in vermiculite + sand (1:1) while, minimum (3.48 cm) in sand + FYM (1:1). At 90 days, maximum (4.41 cm) average length of primary root was observed during 15th-21st of August which was followed by (4.23 cm) during 15th-21st of September while, minimum (3.82 cm) during 15th-21st of June.

4.2.14.2 Interaction Effect of Time and Soil Media for Planting Cuttings on Length of Primary Root (cm) in Guava cv. L-49

The data presented in Table 29 on interaction effect of time and soil media for guava propagation through cuttings during 15th-21st of August showed maximum (4.93 cm) length of primary root in treatment combination Vermiculite + sand + FYM (1:1:1) which was followed by (4.68 cm) in Vermiculite + sand (1:1) while, minimum (3.81 cm) in sand + FYM (1:1). Similarly, during 15th-21st of September maximum (4.76 cm) length of primary root was recorded in Vermiculite + sand + FYM (1:1:1) which followed by (4.60 cm) in Vermiculite + sand (1:1) while, minimum (3.60 cm) in sand + FYM (1:1).

Data pertaining to interaction effect of time and soil media for guava propagation through cuttings during 15th-21st of July recorded maximum (4.60 cm) length of primary root in Vermiculite + sand + FYM (1:1:1) which was followed by (4.37 cm) in Vermiculite + sand (1:1) while, minimum (3.31 cm) in sand + FYM (1:1). Similarly, during 15th-21st of June maximum (4.48 cm) length of primary root was recorded in Vermiculite + sand + FYM (1:1:1) which was followed by (4.00 cm) in Vermiculite + sand (1:1) while, minimum (3.28 cm) in sand + FYM (1:1).

4.2.15 Average Root Fresh Weight (g)

The data pertaining to influence of time and soil media for guava propagation through cuttings on average root fresh weight (g) at 90 days presented in Table 30 showed significant differences among all the treatments.

Table 30: Influence of Soil Media, Time of Planting Cuttings and their Interaction Effect on Average Root Fresh Weight (g) of Guava Cuttings cv. L-49 after 90 Days

Time of Planting	*Cuttings Soil Media*				
	June 15th-21st	*July 15th-21st*	*Aug 15th-21st*	*Sep 15th-21st*	*Mean*
Vermiculite + Sand	1.05	1.07	1.16	1.12	1.10
Perlite + Sand	1.19	1.22	1.31	1.27	1.24
Sand + FYM	1.03	1.06	1.14	1.12	1.09
Vermiculite + Sand + FYM	1.06	1.09	1.17	1.13	1.11
Perlite + Sand+ FYM	1.23	1.27	1.31	1.29	1.27
Mean	**1.11**	**1.14**	**1.21**	**1.18**	

Factors	*S.E. m (±)*	*C.D (0.05)*
Soil media	0.01	0.01
Time of planting cuttings	0.01	0.01
Media x Time	**0.01**	**0.01**

4.2.15.1 Influence of Time and Soil Media for Planting Cuttings on Average Root Fresh Weight (g) in Guava cv. L-49

The data presented in Table 30 showed maximum (1.27 g) average root fresh weight in treatment combination perlite + sand + FYM (1:1:1), which was followed by (1.24 g) in perlite + sand (1:1) while, minimum (1.09 g) in sand + FYM (1:1). At 90 days, maximum (1.21 g) average root fresh weight was recorded during 15th-21st of August which was followed by (1.18 g) during 15th-21st of September while, minimum (1.11 g) during 15th-21st of June.

4.2.15.2 Interaction Effect of Time and Soil Media for Planting Cuttings on Average Root Fresh Weight (g) in Guava cv. L-49

The data presented in Table 30 on interaction effect of time and soil media for guava propagation through cuttings during 15th-21st of August showed maximum (1.31 g) root fresh weight in treatment combination perlite + sand + FYM (1:1:1) and perlite + sand(1:1) while minimum (1.14 g) in sand + FYM (1:1). Similarly, during 15th-21st of September maximum (1.29 g) root fresh weight was recorded in perlite + sand + FYM (1:1:1) which was followed by (1.27 g) in perlite + sand (1:1) while, minimum (1.12 g) in sand + FYM (1:1).

Data pertaining to interaction effect of time and soil media for planting cuttings during 15th-21st of July showed maximum (1.27 g) root fresh weight in perlite + sand + FYM (1:1:1) which was followed by (1.22 g) in perlite + sand (1:1) while, minimum

(1.06 g) in sand + FYM (1:1). Similarly, during 15th-21st of June maximum (1.23 g) root fresh weight was recorded in perlite + sand + FYM (1:1:1) which was followed by (1.19 g) in perlite + sand (1:1) while, minimum (1.03 g) in sand + FYM (1:1).

4.2.16 Average Root Dry Weight (mg)

The data pertaining to influence of time and soil media for guava propagation through cuttings on average root dry weight (mg) at 90 days presented in Table 31 showed significant differences among all the treatments.

Table 31: Influence of Soil Media, Time of Planting Cuttings and their Interaction Effect on Average Root Dry Weight (mg) of Guava Cuttings cv. L-49 after 90 Days

Time of Planting	*Cuttings Soil Media*				
	June 15th-21st	*July 15th-21st*	*Aug 15th-21st*	*Sep 15th-21st*	*Mean*
Vermiculite + Sand	44.67	45.00	50.69	48.70	47.26
Perlite + Sand	48.01	52.68	58.34	55.35	53.84
Sand + FYM	37.34	36.69	45.02	40.68	39.93
Vermiculite + Sand + FYM	46.31	47.66	55.35	50.67	50.10
Perlite + Sand+ FYM	55.02	62.68	67.70	64.70	62.52
Mean	**46.35**	**48.94**	**55.62**	**52.02**	

Factors	*S.E. m (±)*	*C.D (0.05)*
Soil media	0.30	0.85
Time of planting cuttings	0.27	0.76
Media x Time	**0.60**	**1.71**

4.2.16.1 Influence of Time and Soil Media for Planting Cuttings on Average Root Dry Weight (mg) in Guava cv. L-49

The data presented in Table 31 showed maximum (62.52 mg) average root dry weight in treatment combination perlite + sand + FYM (1:1:1) which was followed by (53.84 mg) in perlite + sand (1:1) while, minimum (39.93 mg) in sand + FYM (1:1). At 90 days, maximum (55.62 mg) average root dry weight was observed during 15th-21st of August which was followed by (52.02 mg) during 15th-21st of September while, minimum (46.35 mg) during 15th-21st of June.

4.2.16.2 Interaction Effect of Time and Soil Media for Planting Cuttings on Average Root Dry Weight (mg) in Guava cv. L-49

The data presented in Table 31 on interaction effect of time and soil media for guava propagation through cuttings during 15th-21st of August recorded maximum (67.70 mg) root dry weight in treatment combination perlite + sand + FYM (1:1:1) which was followed by (58.34 mg) in perlite + sand (1:1) while minimum (45.02 mg) in sand + FYM (1:1). Similarly, during 15th-21st of September maximum (64.70 mg) root dry weight was recorded in perlite + sand + FYM (1:1:1) which was followed by (55.35 mg) in perlite + sand (1:1) while, minimum (40.68 mg) in sand + FYM (1:1).

Data pertaining to interaction effect of time and soil media for guava propagation through cuttings during 15th-21st of July recorded maximum (62.68 mg) root dry weight in perlite + sand + FYM (1:1:1) which was followed by (52.68 mg) in perlite + sand (1:1) while, minimum (36.69 mg) in sand + FYM (1:1). Similarly, during 15th-21st of June maximum (55.02 mg) root dry weight was recorded in perlite + sand + FYM (1:1:1) which was followed by (48.01 mg) in perlite + sand (1:1) while, minimum (37.34 mg) in sand + FYM (1:1).

4.2.17 Chlorophyll Percentage

The data pertaining to influence of time and soil media for guava propagation through cuttings on chlorophyll percentage at 90 days presented in Table 32 showed significant differences among all the treatments.

Table 32: Influence of Soil Media, Time of Planting Cuttings and their Interaction Effect on Chlorophyll Percentage of Guava Cuttings cv. L-49 after 90 Days

Time of Planting	*Cuttings Soil Media*				
	June 15th-21st	*July 15th-21st*	*Aug 15th-21st*	*Sep 15th-21st*	*Mean*
Vermiculite + Sand	2.34 (0.17)	2.56 (0.21)	2.78 (0.24)	2.67 (0.22)	2.59 (0.21)
Perlite + Sand	1.95 (0.13)	2.09 (0.14)	2.64 (0.22)	2.43 (0.19)	2.27 (0.17)
Sand + FYM	1.61 (9.09)	2.02 (0.13)	2.50 (0.20)	2.41 (0.19)	2.24 (0.15)
Vermiculite + Sand + FYM	2.43 (0.19)	2.61 (0.21)	2.86 (0.25)	2.74 (0.23)	2.66 (0.22)
Perlite + Sand+ FYM	2.12 (0.14)	2.41 (0.18)	2.74 (0.23)	2.61 (0.21)	2.47 (0.19)
Mean	2.09 (0.14)	2.34 (0.17)	2.70 (0.23)	2.57 (0.21)	
Factors		*S.E. m (±)*		*C.D (0.05)*	
Soil media		0.01		0.01	
Time of planting cuttings		0.01		0.01	
Media x Time		**0.01**		**0.02**	

Note: Figures in parenthesis indicate observed values and others are transformed values.

4.2.17.1 Influence of Time and Soil Media for Planting Cuttings on Chlorophyll Percentage in Guava cv. L-49

The data presented in Table 32 showed maximum (0.22 per cent) average chlorophyll percentage in treatment combination vermiculite + sand + FYM (1:1:1), which was at par with (0.21 per cent) in vermiculite + sand (1:1) while, minimum (0.15 per cent) in sand + FYM (1:1). At 90 days, maximum (0.23 per cent) average chlorophyll percentage was observed during 15th-21st of August which was followed by (0.21 per cent) during 15th-21st of September while, minimum (0.14 per cent) during 15th-21st of June.

4.2.17.2 Interaction Effect of Time and Soil Media for Planting Cuttings on Chlorophyll Percentage in Guava cv. L-49

The data presented in Table 32 on interaction effect of time and soil media for guava propagation through cuttings during 15th-21st of August recorded maximum (0.25 per cent) chlorophyll percentage in treatment combination Vermiculite + sand + FYM (1:1:1) which was at par with (0.24 per cent) in Vermiculite + sand (1:1) and (0.23 per cent) in perlite + sand + FYM (1:1:1) while, minimum (0.20 per cent) in sand + FYM (1:1). Similarly, during 15th-21st of September maximum (0.23 per cent) chlorophyll percentage was recorded in Vermiculite + sand + FYM (1:1:1) which was at par with (0.22 per cent) in Vermiculite + sand (1:1) and (0.21 per cent) in perlite + sand + FYM (1:1:1) while, minimum (0.19 per cent) in sand + FYM (1:1) and perlite + sand (1:1).

Data pertaining to interaction effect of time and soil media for guava propagation through cuttings during 15th-21st of July recorded maximum (0.21 per cent) chlorophyll percentage in Vermiculite + sand + FYM (1:1:1) and Vermiculite + sand (1:1) while, minimum (0.13 per cent) in sand + FYM (1:1). Similarly, during 15th-21st of June maximum (0.19 per cent) chlorophyll percentage was recorded in Vermiculite + sand + FYM (1:1:1) which was at par with (0.17 per cent) in Vermiculite + sand (1:1) while, minimum (0.09 per cent) in sand + FYM (1:1).

Chapter 5

Discussion

Guava (*Psidium guajava* L.) is highly cross pollinated crop. Consequently, being heterozygous in nature, it does not breed true to type plants through seeds. Need for having vegetatively propagated plants in different fruit crops is understood by all for quality fruit production. Replacement of propagation method by cutting, air layering, patch budding and wedge grafting has been tried with the success in each being influenced predominantly by season of propagation. Adoption of any such method in a place like Jammu which has great potential for guava production, calls for standardization of the methods to determine their comparative performance and to find out best season of propagation along with best combination of soil media for planting of guava cuttings. Therefore, in this direction present investigation was carried out to study "Propagation in guava (*Psidium guajava* L.) cv. L-49 under Jammu sub-tropics"

In this chapter the significant experimental findings during the course of investigation have been discussed to offer possible explanations and evidences with a view to find out the cause and effect relationship among different treatments with regards to various attributes studies which are presented as under:

5.1 Influence of Time and Propagation Methods on per cent Success in Guava

Among all the methods of propagation patch budding and wedge grafting behaved almost uniformly as far as per cent success was concerned. However, after 90 days of propagation highest (92.07 per cent) per cent success was recorded in patch budding during 15th-21st of August where as during 15th-21st of September the highest per cent success 87.61 per cent was observed in patch budding which was at par with (86.57 per cent) in wedge grafting. Superiority of patch budding over other methods might be due to the larger bark and cambium tissues in patch which come in contact easily between stock and scion after budding operation. Highest success in patch budding is in accordance with the findings of Kumar *et al.*, 2007. The findings showing success in wedge grafting are in agreement with Gurjar *et al.*, 2012 where they observed that wedge grafting success ranged from 80-90 per cent when grafting operation was performed during September to March. Ahmad *et al.*, 2007 also observed maximum bud-take in walnut when the patch budding was performed on 30 July followed by 30 August. This might be due to rapid complete union of xylem and cambium tissue of the scion and rootstock or due to much closer matching of the scion tissue to the rootstock stem during this period that helps in callus tissue differentiation into cambium tissue (Hartmann *et al.*, 1997., Janick, 1982). The minimum or below average percentage of successful bud-take was recorded in the case of budding performed during 15th-21st of June, the low bud-take percentage during this period might be due to immature bud wood and low sap flow.

5.2 Influence of Time and Propagation Methods on Growth Characteristics of Guava

Interaction effect of time and propagation methods significantly influenced the number of days to sprout and plant height in guava. Minimum (11.75) number of days to sprout and maximum (61.89 cm) plant height was observed in air layering during 15th-21st of August which was followed by (12.76) days taken to sprout and plant height (58.47 cm) in air layering during 15th-21st of September. Early sprouting and maximum increase in plant height in air layered plants as compared to plants propagated with other methods might be due to the reason that in air layering there is no callus formation and bud union as in wedge grafting and patch budding. The results are in accordance with Rymbai and Reddy, 2010 in air layering of guava, where they demonstrated that root characters had progressively improved from 15th June to 15th August, suggesting that the rooting was affected due to environmental conditions, steady increase in relative humidity from June to August with temperature approaching down from high temperature of summer to moderate temperature of rainy and autumn season made the conditions congenial for growth and development of plant in month of August.

After 90 days of propagation maximum (98.00 per cent) sprouting was obtained in patch budding during 15th-21st of August which was at par with (93.68 per cent) patch budding during 15th-21st of September and in wedge grafting (91.90 per cent) during 15th-21st of September. These results are in line with the findings of Negi *et al.*, 2010 where they reported that patch budding gave the highest mean bud sprouting in anola among other methods of propagation. The possible cause for supremacy of patch budding might be that it is easy to remove the bark in the form of rectangular patch. Kumar *et al.*, 2005 also reported that maximum bud sprout in ber (*Zizyphus mauritiana*) was observed during beginning of August and mid-August as compared to the budding done during the month of September and October. Good results of sprouting in wedge grafting are in consonance with that of Visen *et al.*, 2010 where they reported that highest sprouting was observed by use of poly cap irrespective of scion cultivars and month of grafting due to creation of high humidity around graft-scions which reduced the desiccation of active tissue of scion bud as compared with open conditions. Seasons of propagation play an important role in per cent sprouting. Maximum sprouting was observed during 15th-21st August and during 15th-21st of September when temperature was 32 °C to 33 °C and high relative humidity, which might be due to early callus formation and proliferation at bud union. The lower success during June is also due non availability of good bud wood because of fruit set on the trees.

Wedge grafting during 15th-21st of August was found to be excellent after 180 days of propagation with respect to stem thickness (1.34 cm) which was followed by stem thickness (1.33 cm) in patch budding. These results are in agreement with the findings of Somkuwar *et al.*, 2009 where they reported that wedge grafting during 15 August in grape cv. Tas-A-Ganesh as scion resulted in thickest shoot.

After 180 days of propagation number of shoots per plant (8.97) and number of leaves per plant (22.39) were found maximum in wedge grafted plants during 15th-21st of August which was followed by number of shoots (6.88) and number of leaves (19.77) in patch budding during 15th-21st of August. The maximum number of shoots and leaves in wedge grafting might be due to presence of 3-4 buds on scionwood used for wedge grafting instead of single bud on patch used for patch budding. Visen *et al.*, 2010 and Singh *et al.*, 2011 also reported that number of bud sticks on scion stick 3-4 buds was all direction ideal for wedge grafting. Findings showing maximum number of shoots and leaves during 15th-21st of August are in line with the results of Gurjar and Singh, 2012 in anola, where they observed that during rainy season well matured rootstock favoured with high atmospheric humidity along with fairly high temperature, is found congenial for rapid callus production that ensures formation of an early and strong union between stock and scion.

Response of wedge grafting after 90 days was also found to be best in terms of average leaf area (18.28 cm^2), leaf fresh weight (0.34 g) and leaf dry weight (25.96 gm). Chlorophyll percentage (0.19 per cent) was also found maximum in wedge grafted plants and this is in conformation with findings of Bao *et al.*, 2012 where they reported that chlorophyll content directly affected the survival rate of propagated plants. Singh *et al.*, 2007 also reported that wedge grafting performed during August and September gave good success as compared to wedge grafting performed during June and July. Saroj *et al.* (2000), Tiwari and Bajapai (2000) has also confirmed the same findings in aonla and reported that highest success of graft is possible in August (85 per cent).

5.3 Influence of Soil Media and Time of Propagation on Guava

Vermiculite + sand + FYM (1:1:1) treatment combinations of soil media used for planting guava cuttings was found to be best among all treatments tried. Where, after 90 days of planting highest (78.69 per cent) per cent success was recorded in vermiculite + sand + FYM (1:1:1) during 15th-21st of August which was at par with (77.02 per cent) in vermiculite + sand (1:1) and (76.01 per cent) in perlite + sand + FYM (1:1:1). Results showing the superiority of media containing vermiculite are in consonance with the observation of Hossenini *et al.*, 2004 where they reported that vermiculite has been recorded a positive affect on the rooting of Iranian cultivars when used as a medium component for cuttings. Okao *et al.*, 2012 also stated that cuttings rooted in vermiculite produced significantly higher number of rooted cuttings than those rooted in other substrates.

The time of planting cuttings, soil media and their interaction significantly influenced the number of days taken for sprout. Minimum (8.95) number of days taken to sprout was observed in treatment combination perlite + sand + FYM (1:1:1) during 15th-21st of August. These findings are in conformity with that of Conover and Joiner, 1963 where they reported that perlite having pH 6.7 and vermiculite with pH 7.5 may be high enough to suppress initiation and growth of roots. Vermiculite stuck together and held a thin layer of moisture on the surface, resulting in poor aeration of the media.

Response of vermiculite + sand + FYM (1:1:1) during 15th-21st of August was also found to be best after 180 days in terms of number of shoots per plant (10.42) which was followed by (10.19) shoots in vermiculite + sand + FYM (1:1:1) during 15th-21st of September. These findings are in line with the results of Bethke, 2007 where he reported that vermiculite is having high nutrient holding capacity than perlite and contributes potassium and magnesium to the media which resulted in more shoots.

Vermiculite + sand + FYM (1:1:1) combinations of soil media used for planting guava cuttings showed maximum increase in plant height (26.49 cm) which was followed by (23.19 cm) cuttings planted in perlite + sand + FYM (1:1:1) during 15th-21st of August after 180 days of planting. These results are in accordance with the findings of Cros *et al.*, 2007 where they reported that substrate containing vermiculite exhibited more plant height as compared to plant height in perlite.

The time of planting cuttings, soil media and their interaction significantly influenced the number of leaves. Maximum number of leaves (25.08) was recorded in vermiculite + sand + FYM (1:1:1) among other media during 15th-21st of August which was followed by (24.99) leaves in vermiculite + sand + FYM (1:1:1) during 15th-21st of September after 180 days of planting. The findings are in conformity with the results of Ali (2011) where he reported that number of leaves per plant was maximum in substrate containing vermiculite and less number of leaves per plant was recorded in substrate containing perlite in dahlia plants.

Treatment combination vermiculite + sand + FYM (1:1:1) resulted in higher subsequent growth with respect to leaf fresh weight (0.43 g) and leaf dry weight (27.00 mg) during 15th-21st of August after 90 days which was followed by leaf fresh weight (0.42 g) and leaf dry weight (26.90 mg) in vermiculite + sand (1:1) during 15th-21st of August. These findings are in consonance with Cros *et al.* (2007) where they reported that substrate containing vermiculite resulted in more fresh weight of plants as compared to substrate containing perlite in common purslane which resulted in more leaf dry weight of plants grown in media containing vermiculite.

Response of cuttings planted in vermiculite + sand + FYM (1:1:1) among other media during 15th-21st of August, was found to be excellent after 90 days of planting with respect to average leaf area (22.17 cm^2) which was followed by (20.18 cm^2) in vermiculite + sand + FYM (1:1:1) during 15th-21st of September. The findings are in agreement with the results of Ali (2011) where he observed maximum leaf area index in plants grown in substrate containing vermiculite and less leaf area index of leaves in plants grown in substrate containing perlite.

Treatment combination vermiculite + sand + FYM (1:1:1) after 90 days resulted in maximum plant fresh weight (4.78 g) and plant dry weight (0.23 g) during 15th-21st of August which was followed by (4.72 g) plant fresh weight and (0.19 g) plant dry weight in vermiculite + sand + FYM (1:1:1) during 15th-21st of September. Above findings are in accordance with the results of Martinez-Medina *et al.*, 2009 where they reported that plants treated with the bentonite-vermiculite formulation of *Trichoderma harzianum* showed significant increase in fresh and dry weights compared to non formulated plants.

The time of planting cuttings, soil media and their interaction significantly influenced the number of primary and secondary roots. Vermiculite + sand + FYM (1:1:1) among other media during 15th-21st of August, was found to be excellent after 90 days of planting with respect to number of primary roots (5.26) and secondary roots (9.44) which was followed by (5.20) primary roots and (9.35) secondary roots in Vermiculite + sand + FYM (1:1:1) during 15th-21st of September. These findings are in line with the results of Copes (1977) where they observed that use of vermiculite promoted lateral branches but appeared to have little effect on root thickness, while high proportions of perlite usually resulted in short and not well branched roots. He also reported that media containing only perlite or perlite + sand dried out rapidly and needed more frequent and heavy watering than others. It was almost impossible to overwater the perlite –sand media. Vermiculite was somewhat intermediate in water retention characteristics.

Response of cuttings planted in treatment combination perlite + sand + FYM (1:1:1) during 15th-21st of August, was found to be excellent after 90 days of planting with respect to root fresh weight (1.31 g) and root dry weight (67.7 mg). These findings are in conformity with the results of Majdi *et al.* (2012) where they reported that substrate containing perlite showed maximum root weight then substrate containing vermiculite. Results of Copes (1977) also showed that perlite based media have thicker roots whereas vermiculite has no effect on root thickness.

Vermiculite + sand + FYM (1:1:1) treatment combinations of soil media used for planting guava cuttings was found to be best among all treatments tried. Where, after 90 days of planting maximum (0.25 per cent) chlorophyll percentage was recorded in vermiculite + sand + FYM (1:1:1) during 15th-21st of August which was at par with (0.23 per cent) in vermiculite + sand + FYM (1:1:1) during 15th-21st of September. These findings are in agreement with the results of Martinez-Medina *et al.*, 2009 where he reported that plants treated with the bentonite-vermiculite showed significant increase in chlorophyll content.

Findings showing poor performance of substrate containing perlite as compared to vermiculite are in agreement with Bruckner (1979) where he reported that relative balance of air and water within a soil's pore space is critical to plant growth. Perlite is generally unsatisfactory for the production of plants in containers. This is primarily because perlite does not provide the aeration, drainage and moisture in good balance at low tensions required. Ors and Anapali (2010) also reported that perlite is generally unsatisfactory for the production of plants in containers.

Response of cuttings was found to be best during 15th-21st of August is in consonance with the findings of Gautam *et al.* (2010) who reported that maximum rooting in cuttings of guava took place in rainy reason. However, relatively poor root formation recorded during extremely hot and cold conditions was due to low activity of cambium to proliferate in unfavourable and adverse environmental conditions (Goel and Behl, 1994). Hafeez ur Rehman *et al.* (1990) also reported maximum success in cuttings from June 15 to August 15. Maximum success during 15th-21st of August might be due to favourable temperature, high relative humidity, long sun-shine hours and low evaporation rate like climatic conditions which are congenial for growth and development of plant in month of August.

Chapter 6

Summary and Conclusions

Propagation of plants may be defined as the control reproduction of plants by man to perpetuate released individuals or group of plants which have specific value to him. Plant propagation involves the application of specific biological principles and concepts in multiplication of plants for useful purposes. Therefore, the present investigation entitled "Propagation studies in guava (*Psidium guajava* L.) cv. L-49 under Jammu sub-tropics" was conducted during 2012-13 at Fruit Plant Nursery, Faculty of Agriculture, Sher-e-Kashmir University of Agricultural Sciences and Technology of Jammu, Udheywalla, Jammu. The findings have clearly indicated that there was a positive effect of time and methods of propagation along with the soil media on propagation of guava. The significant findings found during the course of study are summarized below:

6.1 Influence of Time and Methods of Propagation on per cent Success in Guava

Highest per cent success (92.07 per cent) as influenced by time and method of propagation was observed in patch budding during 15th to 21st of August which was at par with per cent success (87.61 per cent) in patch budding and (86.57 per cent) in wedge grafting during 15th to 21st of September.

6.2 Influence of Time and Methods of Propagation on Growth Characteristics of Guava

Number of days taken to sprout was found minimum (11.75) in air layering during 15th to 21st of August as influenced by time and method of propagation.

After 90 days of propagation the highest percentage sprouting (98.00 per cent) was observed in patch budding during 15th to 21st of August.

At 60, 90, 120, 150, 180 days, plant height in propagated guava plants (53.60 cm, 54.80 cm, 57.83 cm, 60.00 cm and 61.89 cm) were found maximum in air layering during 15th to 21st of August.

At 60, 90, 120, 150, 180 days, number of shoots in propagated plants (3.69, 4.07, 4.98, 5.08, and 8.97) were observed maximum in wedge grafting during 15th to 21st of August.

At 60, 90, 120, 150, 180 days, number of leaves in propagated plants (9.19, 15.00, 18.67, 19.44, and 22.39) were recorded maximum in wedge grafting during 15th to 21st of August.

At 60, 90, 120, 150, 180 days, stem thickness of propagated plants (1.16 cm, 1.19 cm, 1.24 cm, 1.31 cm, 1.34 cm) were recorded maximum in wedge grafting during 15th to 21st of August.

After 90 days of propagation, maximum (18.28 cm^2) leaf area was obtained in wedge grafting during 15th to 21st of August.

At 90 days of propagation, maximum (0.34 g) leaf fresh weight was recorded in wedge grafting during 15th to 21st of August.

At 90 days of propagation, maximum (25.96 gm) leaf dry weight was recorded in wedge grafting during 15th to 21st of August.

At 90 days of propagation, maximum (0.19 per cent) chlorophyll percentage was found in wedge grafting during 15th to 21st of August.

6.3 Influence of Time and Soil Media on Propagation in Guava

Highest percentage of rooted cuttings as influenced by time and soil media on propagation of guava (78.69 per cent) was observed during 15th to 21st of August in cuttings planted in soil media of vermiculite + sand + FYM (1:1:1).

Influence of soil media and time of planting cuttings on number of days taken to sprout was found minimum (8.95) in cuttings planted in perlite + sand + FYM (1:1:1) during 15th to 21st of August.

At 60, 90, 120, 150, 180 days of planting, number of shoots in cuttings (4.88, 5.34, 7.80, 8.97 and 10.42) were observed maximum in cuttings planted in media of vermiculite + sand + FYM (1:1:1) during 15th to 21st of August.

At 60, 90, 120, 150, 180 days of planting, plant height (17.57 cm, 21.17 cm, 23.39 cm, 25.38 cm and 26.49 cm) was found highest in cutting planted in soil media of vermiculite + sand + FYM (1:1:1) during 15th to 21st of August.

At 60, 90, 120, 150, 180 days of planting, stem thickness (1.03 cm, 1.07 cm, 1.09 cm, 1.11cm and 1.14cm) were recorded maximum in cuttings planted in soil media of vermiculite + sand + FYM (1:1:1) during 15th to 21st of August.

At 60, 90, 120, 150, 180 day of planting, number of leaves (10.40, 15.88, 18.62, 21.82 and 24.29) were obtained maximum in cuttings planted in soil media of vermiculite + sand + FYM (1:1:1) during 15th to 21st of August.

At 90 days of planting, leaf area (22.17 cm^2) was obtained maximum in cuttings planted in soil media of vermiculite + sand + FYM (1:1:1) during 15th to 21st of August.

At 90 days of planting leaf fresh weight (0.43 g) was found to be maximum in cuttings planted in soil media of vermiculite + sand + FYM (1:1:1) during 15th to 21st of August.

At 90 days of planting leaf dry weight (27.00 mg) was found to be maximum in cuttings planted in soil media of vermiculite + sand + FYM (1:1:1) during 15th to 21st of August.

At 90 days of planting plant fresh weight (4.78 g) was found to be maximum in cuttings planted in soil media vermiculite + sand + FYM (1:1:1) during 15th to 21st of August.

At 90 days of planting plant dry weight (0.23 g) was found to be maximum in cuttings planted in soil media of vermiculite + sand + FYM (1:1:1) during 15th to 21st of August.

At 90 days of planting cuttings number of primary roots per rooted cutting (5.26) was found to be maximum in cuttings planted in soil media of vermiculite + sand + FYM (1:1:1) during 15th to 21st of August.

At 90 days of planting cuttings number of secondary roots per rooted cutting (9.44) was found to be maximum in cuttings planted in soil media of vermiculite + sand + FYM (1:1:1) during 15th to 21st of August.

At 90 days of planting cuttings length of primary root (4.93 cm) was found to be maximum in cuttings planted in soil media of vermiculite + sand + FYM (1:1:1) during 15th to 21st of August.

At 90 days of planting cuttings average root fresh weight (1.31 g) was found to be maximum in cuttings planted in soil media of perlite + sand + FYM (1:1:1) during 15th to 21st of August.

At 90 days of planting cuttings average root dry weight (67.70 gm) was found to be maximum in cuttings planted in soil media of perlite + sand + FYM (1:1:1) during 15th to 21st of August.

At 90 days of planting cuttings chlorophyll percentage (0.25 per cent) was found to be maximum in cuttings planted in soil media of vermiculite + sand + FYM (1:1:1) during 15th to 21st of August.

Conclusions

Based on the experimental results obtained, it may be concluded that patch budding performed during 15th to 21st of August was found to be the best method for guava propagation under Jammu Sub-tropics.

In the second experiment it may be concluded that guava cuttings planted in soil media of vermiculite + sand + FYM (1:1:1) during 15th to 21st of August was found to be best suitable media for guava propagation under Jammu Sub-tropics.

Literature Cited

Abdullah, A.T.M., Hossain, M.A., and Bhuiyan, M.K. 2006. Clonal propagation of guava (*Psidium guajava* Linn.) by stem cutting from mature stockplants. *Journal of Forestry Research,* **17** (4): 301-304 (2006).

Adsule, R. N. and Kadam, S. S. 1995. Guava. In: Salunkhe, D. S. and Kadam, S. S. (eds.). *Hand Book Fruit Science and Technology: Production, Composition, Storage and Processing*. 419-433. Marcel Dekker, I. L. C, New York.

Ahmad, M. F., Iqbal, U. and Khan, A. A. 2007. Response of different environments and date of patch budding on success in walnut. *Indian Journal of Horticulture,* **64** (3): 286-289.

Ali, Y. S. S. 2011. Effect of mixing palm leaves compost (DPLC) with vermiculite, perlite, sand and clay on vegetative growth of dahlia (*Dahlia pinnata*), marigold (*Tagetes erecta*), zinnia (*Zinnia elegans*) and cosmos (*Cosmos bipinnatus*) plants. Department of Plant Production, College of Food Science and Agriculture, King Saud University.

Anonymous, 2012. National Horticulture Board Database, www. nhb.org.

Anonymous, 2012. Area and production under fruit crops. *Directorate of Horticulture.* Jammu and Kashmir.

Athani, S. I., Swamy, G.S.K. and Patil, P. B. 1999. Responses of guava (*Psidium guajava* L.) varieties to air layering. *Advances in Agricultural Research in India,* **11**: 63-65

Aulakh, P. S. 1998. Standardization of patch budding time in guava under rainfed conditions in the lower foothills Shiwalik of Punjab - a note. *Progressive Horticulture,* **30** (3-4): 221-222.

Ayaz, M., Hussain, S.A. and Ali, N. 2004. Effect of paclobutrazol concentration and dipping period on rooting of soft wood cuttings of guava (*Psidium guajava*), *Pakistan Journal of biological sciences,* **7** (1):28-31.

Babu, K.D. and Yadav, D.S. 2007. Genotypic Amenability of guava for patch budding under sub tropical conditions of north east India. *Acta horticulture*, **735**:231-234.

Babu, K.D., De, L.C., Patel, R.K. and Singh, A. 2009. Genotypic amenability of guava for patch budding. *Indian Journal of Horticulture*, **66** (2): 264-266.

Bahuguna, V. K. and Pyarelal. 1990. To study the effect of environment and different soil mixture on germination of *Acacia nilotica* seed at nursery stage. *Indian Forestry*, **116**: 474- 478.

Bao, R., Yin, P., Dai, J., Guo, B. and Wei, Y. 2012. Effects of different media on the transplantation of *Huperzi serrata* (Thunb.) Trev. *African Journal of Agricultural Research*, **7** (20): 3045-3048.

Bethke, C. L. 2007. Rice hulls vs perlite and vermiculite as a growing media component. Perlite Institute, Inc.

Bhagat, B. K, Singh, C. and Chaudhary, B. M. 1999. Effect of growth substances on rooting and survival of air layers of guava (*Psidium guajava* L.) cv. Sardar. *Orissa Journal of Horticulture*, **27** (2): 72-75.

Bruckner, U. 1979. Physical properties of different potting media and substrate mixtures- especially air-and water capacity, *Acta Horticulture*, **450**: 263-270.

Caldwell, J. D., Coston, D. C. and Brock, K. H. 1988. Rooting of semi-hardwood 'Hayward' kiwifruit cuttings. *HortScience*, **23**(4): 714-717.

Canover and Joiner, 1963. Rooting response of Pittosporium tobira variegatum as affected by 3- indolebutyric acid, rooting media and age of wood. *Florida Agricultural Experimental Stations Journal*, series no. 1781.

Copes, D.L. 1977. Influence of rooting media on rootstructure and rooting percentage of douglas –fir cuttings. *Silvae Genetica*, **26**: 102-106.

Costa, E., Rodrigues, E. T., Santos, L. C. R. D., Vieira, L. C. R. and Ferreira, M. D. 2008. Protected environments on the development of passion fruit and papaya seedlings in the southern pantanal matogrossense. *CIGR – International Conference of Agricultural Engineering XXXVII Congresso Brasileiro de Engenharia Agricola*.

Cros, V., Martinez- Sanchez, J. J., Franco, J. A. 2007. Good yields of common purslane with a high fatty acid content can be obtained in a peat-based floating system. *Horticulture Technology*, **17**:14-20.

Fabbri, A., Bartolini, G., Lambardi, M. and Kailia, S. 2004. Olive propagation manual, Landlinks Press, Colling-wood, 141 pp.

Gautam, N.N., Singh, K., Singh, B., Seal, S., Goel, A. and Goel, V.L. 2010. Studies on clonal multiplication of Guava (*Psidium guajava* L.) through cutting under controlled conditions. *Australian Journal of Crop Science*, **4** (9):666-669.

Ghosh, S. N. and Ranjan, T. 2005. Air layering in guava in red laterite zone of West Bengal. *Horticultural Journal*, **18** (2): 91-93.

Goel, V. L. and Behl, H. M. 1994. Cloning of selected genotypes of promising tree species for sodic soil sites. *Journal of Indian Botanical Society*, **73**: 255-258.

Gupta, M.R. and Mehrotra, N.K. 1985. Propagation studies in guava (*Psidium guajava* L.) cv. Allahabad Safeda. *Journal of Research of Punjab Agricultural University,* **22**: 267-69.

Gurjar, P. S. and Singh, R. 2012. Performance of wedge grafting in anola at polyhouse and open field conditions. *Environment and Ecology,* **30** (3): 531-536.

Gurjar, P.S., Singh, R., Maskar, S.B., Singh, N. and Choubey, R. 2012. Propagation of guava by wedge grafting under polyhouse and open field cincitions. *Plant Archives* **(12)**2:827-832.

Hafeez, U. R., Khan, M. A. and Khokar, K. 1990. Effect of season on rooting ability of tip cutting of guava treated with paclobutrazol, *Indian Journal of Agricultural Sciences,* **61** (6): 404-406.

Hartmann, H. P., Kester, D. S., Davies, F. T. and Geneve, R. L. 1997. *Plant Propagation –Principles and Practices* (6th edition). Prentice Hall of India Pvt. Ltd., New Delhi.

Hosseini, S.M., Sadeghi H., Esmati, A., Nourmohammadi, Z., Keshavarz, M. A., Hosseinimaziani, M. 2004. Effect of media on rooting cuttings of four olive cultivars. *5th International Symposium on Olive Growing,* October 2-9, Ýzmir, Turkey. pp. 234 (Abstract).

Isfendiyaroglu, M., Ozeker, E nd Baser,S. 2009. Rooting of 'Ayvalik' olive cuttings in different media. *Spanish Journal of Agricultural Research,* **7** (1). 165-172.

Jackson, M. L. 1967. *Soil Chemical Analysis,* pp 375. *Prentice Hall Inc, New York, U. S. A.*

Jackson, M. L. 1973. *Soil Chemical Analysis,* pp 183-184. Prentice Hall of India, Private Limited, New Delhi.

Janick, J. 1982. *Horticultural Science* (3rd edition). Surjeet Publications, Delhi.

Kareem, A., Jaskani, M.A., Fatima, B and Sadia, B. 2013. Clonal multiplication of guava through softwood cuttings under mist conditions. *Pakistan Journal of Agricultural Science,* **50** (1), 23-27; 2013.

Kaundal, G. S., Gill, S. S. and Minhas, P. P. 1987. Budding techniques in colonal propagation of guava, *Horticulture Journal,* **3**: 37-42.

Khattak, M. S., Malik, M. N. and Khan, M.A. 1983. *In vivo* propagation of guava (*Psidium guajava* L.) through T-grafting. *Pakistan Journal of Agriculture Sciences,* **30** (3): 287-292.

Kilany, O. and Gabr, M. F. 1986. Propagation of seedless guava trees by cuttings. *Annals of Agricultural Science, Moshtohor,* **24** (2): 953-964.

Kumar, G., Dhaliwal, H. S., Aulakh, P. S. and Baidwan, R. P. S. 2005. Standardization of time of budding in ber (*Ziziphus mauritiana* Lamk) under rainfed conditions in lower shivaliks of Punjab. *Environment and Ecology,* **23S** (Special 4): 654-656.

Kumar, K., Aulakh, P. S. and Baidwan, R. P. S. 2007. Standardization of time of budding in guava (*Psidium guajava* L.) under lower shivaliks conditions of Punjab. *Haryana Journal of horticultural Sciences,* **36** (1-2): 61-62.

Kumar, S. and Syamal, M. M. 2005. Effect of etiolation and plant growth substances on rooting and survival of air-layers of guava. *Indian Journal of Horticulture,* **62** (3): 290-292.

Loach, K. 1988. Controlling environmental conditions to improve adventitious rooting. In: Adventitious root formation in cuttings (Davis, T. D., Haissig, B. E. and Sankhla, N., eds.). Dioscorides Press, Portland, Oregon, pp. 248-279.

Majdi, Y., Ahmadizadeh, M. and Ebrahimi, R. 2012. Effect of different substrates on growth indices and yield of green peppers at hydroponic cultivate. *Current Research Journal of Biological Sciences,* **4** (4): 496-499.

Manan, A., Khan. M.A., Ahmad, W. and Sattar, A. (2002).Clonal propagation of guava (Psidium guajava L.). *International journal of agriculture and biology,* (1):143-144.

Marinho, C.S. 2009. Guava propagation through minicuttings. *Revista Brasileria De Fruticultura,* **31** (2): 607-611.

Martinez-Medina, A., Roldan, A. and Pascual, J. A. 2009. Performance of a *Trichoderma harzinum* bentonite-vermiculite formulation against fusarium wilt in seedling nursery melon plants. *Horticulture Science,* **44** (7): 2025-2027.

Mehrotra, N. K. and Gupta, M. R. 1984. A note on vegetative propagation of guava cv. Luckhnow-49. *Haryana Journal of horticultural Sciences,***13**: 135-136.

Mir, K. and Kumar, A. 2011. Effect of different methods, time and environmental conditions on grafting in walnut. International Journal of Farm Sciences **1**(2): 17-22.

Negi, R. S., Baghel, B. S., Gupta, A. K. and Singh, Y. K. 2010. Standardization of method of orchard establishment and propagation in anola (*Emblica officinalis* Gaertn) for rehabilitation of degraded pasture/grazing lands. *Annals of Horticulture,* **3** (1): 39-46.

Okao, M., Malinga, M., Okia, C. A. and Okullo, J. B. L. 2012. Vegetative propagation of Vitellaria paradoxa by stem cuttings: Effects of rooting substrate and planting technique. *3rd Ruforum Binnial Meeting,* 24-25 September, Entebbe, Uganda.

Olsen, S. R., Cole, C. W., Watenable, P. S. and Dean, L. A. 1954. Estimation of average phosphorus in soil by extraction with $NaHCO_3$. *United States Department of Agriculture Circular,* **939**: 19.

Ors, S. and Anapali, O. 2010. Effect of soil addition on physical properties of perlite based media and strawberry cv. Camarosa plant growth. *Scientific Research Essays,* **5** (22):3430-3433.

Pandey, I. C., Upadhyay, N. P and Prasad, R. S. 1979. Vegetative propagation of guava. *Indian Horticulture,* **24** (2): 3-4.

Panse, V. G. and Sukhatme, P.V. 2000. Statistical Methods for Agricultural Workers. *Publication and Information Division of ICAR,* New Delhi.

Patel, N. B. and Pasaliya, Y. M. 1995. Effect of plant growth regulators and advanced ringing on rooting in air- layering of guava (*Psidium guajava* L.) cv. L-49. *Recent Horticulture*, **2** (2): 33-36.

Patel, R.K., Yadav, D.S., Singh, A. and Yadav, R.M. 2007. Performance of patch budding on different cultivars/hybrids of guava under mid hills of Meghalaya. *Acta Horticulture*,**735**: 189-191.

Patel, R.K., Yadav., D.S. Yadav, R.M. and Prasad, R.M. and Patel, K.K 2005. Performance of patch budding on different cultivars of guava under mid hill altitude of Meghalaya. *Orissa Journal of horticultiure*, **33** (1): 1-3.

Pathak, R. K. and Ojha, C.M. 1993. Genetic resources of guava. In: Advance in Horticulture [C]. Vol. I, Fruit crops, Part I, K.L. Chadha and O. P. Pareek (eds), Malhotra Publishing House, New Delhi, 143-147.

Patil, K. B. 2004. Vegetative propagation techniques in anola (*Emblica officinalis*). *Journal of Soils and Crops*, **14**: 194-195.

Piper, C. S. 1966. *Soil and Plant Analysis*, pp 47-49. Hans Publications, Bombay, India.

Pyarelal, K. and Karnataka, K.C., 1993. Effect of orientation of seed sowing and soil mixture on germination behaviour of *Quercus leucotrichophore*. *Indian Forestry*, **119**: 122-125.

Rahman, H. U., Khan, M. A., Khokhar, K. M. and Laghan, M. H. 1991. Effect of season on rooting ability of tip cuttings of guava (*Psidium guajava*) treated with paclobutrazol. *Indian Journal of Agricultural Sciences*, **61**(6): 404-406.

Rahman, H.U., Khan, M.A., Niazi, Z.M. and Khan, D.A. 1988. Rooting of different type of guava stem cuttings using growth regulator. *Pakistan Journal of Agriculture Research*, **9** (3): 263-365.

Rahman, N., Nabi, T. G. and Jan, T. 2003. Effect of different growth-regulators and types of cuttings on rooting of guava (*Psidium guajava L.*). *Science Vision*, **9** (1-2): 1-5.

Rao, Y.R., Kaul, G.L. and Suryanarayana, V. 1984. Studies on vegetative propagation of guava (*Psidium guajava*). *Andhra Agricultural Journal*, **31** (4): 277-281.

Rymbai, H., Reddy, G. S. and K. C, S. 2012. Effect of cocopeat and sphagnum moss on guava air layers and plantlets under open and polyhouse nursery. Agricultural Science Digest, 32 (3): 241-243.

Rymbai, H. and Reddy, G.S. 2010. Effect of IBA, time of layering and rooting media on air layers and plantlets survival under different growing nursery conditions in guava. Indian Journal of Horticulture, 67 (special Issue).

Rymbai, H. and Reddy, G.S.N. 2011. Influence of open field and polyhouse nursery on survival characters of rooted layers in guava (*Psidium guajava* L.). *Life sciences Leaflets*, 21: 996 – 1002.

Samson, J.A., 1986. Tropical Fruits 2nd Edition. Longman Scientific and Tech. Longman and Tech. Longman House Burnet MILL. Harlow. Esex CM 20 2 JE England, 270-275.

Sarkar, A. and Gosh, B. 2006. Air layering in guava cv. L-49 as affected by plant growth regulators, wrappers and time of operation. *Environment and Ecology*, **24S** (3A): 820-823.

Saroj, P. L., Nath, V. and Vashitha, B. B. 2000. Effect of poly containers on germination, seedling vigour, root characters and budding success in anola. *Indian Journal of Horticulture*, **57**: 300-304.

Shanker. S. 1999. Practical manual in horticulture. Balyog Prakashan, 41 pp.

Sharma, R. S., Sharma, T. R. and Sharma, R. C. 1991. Influence of growth regulators and time of operation on rootage of air layering in guava (*Psidium guajava* L.) cv. Allahabad Safeda. *Orissa Journal of Horticulture*, **19** (1-2): 41-45.

Sharma, S., Wali, V.K., Sharma, M., Gupta, V. and Khajuria, S. 2012. Cultivation of micro propagated plants of strawberry (*Fragaria x ananassa* Duch.) cv. Chandler in specific climatic conditions of Jammu (J and K): Hardening and acclimatization *Journal of cell and tissue research*, **12** (3):3373-3376.

Singh, C. P., Singh, R. and Singh, G. 2005. Year around propagation of mango by cleft grafting under controlled conditions. International Conference on Plasticulture and Precision Farming, 17-21 November, 2005, New Delhi pp-71.

Singh, G. 2005. High density planting in guava- application of canopy architecture. ICAR *News* (April-June), **11** (2): 9-10.

Singh, G. and Pandey, D. 1998. Effect of time and method of budding on propagation of guava. (*Psidium guajava* L.). *Annals of Agricultural Research*, **19** (4): 445-448.

Singh, G., Pandey, S. and Singh K. 2011. Vegetative propagation of guava through wedge grafting. *Progressive Horticulture*, **43** (2): 203-210.

Singh, G., Gupta S, Mishra R, Singh G. P. 2005. Wedge grafting in guava- A novel vegetative propagation technique, Publications CISH, Lucknow, 12.

Singh, G., Gupta, S., Mishra, R. and Singh, A. 2007. Technique for rapid multiplication of guava (*Psidium guajava* L.). *Acta Horticulturae*. **735**: 177-183.

Singh, G. and Bajpai, A. 2003. Hi-tech nursery with special reference to fruit crops. In: Precision Farming in Horticulture, Singh, H.P., Singh, Gorakh, Samuel, J. C. and Pathak, R.K. (eds). NCPAH, DAC, MoA, PFDC, CISH, Lucknow, pp. 261- 274.

Singh, J., Sharma, D. P. and Kashyap, R. 1992. Effect of defoliation on survival and growth of guava (*Psidium guajava* L.) layers cv. Seedless. *Advances in Plant Sciences*, **5** (1): 176-179.

Somkuwar, R. G., Satisha, J. and Ramteke, S. D. 2009. Propagation success in relation to time of grafting in Tas-A-Ganesh grapes. *Journal of Maharashtra Agricultural Univerisity*, **34** (1): 113-114.

Subbiah, B. V. and Asija, G. L. 1956. A rapid procedure for the estimation of available nitrogen in soil. *Current Science*, **25**: 259-260.

Sudhakar, K., Mammen, W., Santoshkumar, A.V. and Asokan, P.K. 1995, Effectt of seed size, rooting medium and fertilizers on containerized seedlings on *Liba Pentandra. Indian Forestry*, **12**: 1135-1142.

Sutter, E. G. 2005. Olive cultivars and propagation. In: Olive production manual (Sibbett, G. T. and Ferguson, L., eds.), 2nd ed., University of California, Agriculture and Natural Resources, Publication, 3353, pp. 19-25.

Syamal, M. M., Katiyar, R. and Joshi, M. 2012. Performance of wedge grafting in guava under different growing conditions. *Indian Journal of Horticulture,* **69** (3):424-427.

Tiwari, R. K. and Bajapai, C. K. 2000. propagation of anola (*Emblica officinalis*) through grafting in polybags. *Indian Journal of Agricultural Sciences,* **72**: 353-354.

Ullah, T., Wazir, F.U., Ahmad, M., Analoui, F., Khan, M.U. and Ahmad, M. 2005. A break through in guava (*Psidium guajava* L.) propagation from cutting. *Asian Journal of plant sciences,* **4** (3): 238-243.

Visen, A., Singh, J.N. and Singh, S.P. 2010. Standardization of wedge grafting in guava under North Indian plains. *Indian Journal of Horticulture,* ***67****: 111-114.*

Wahab, F., Nabi, G., Ali, Nawab and Shah, M. 2001. Rooting response of semi-hardwood cuttings of guava (*Psidium guajava* L.) to various concentrations of different auxins. *Online Journal of Biological Sciences,* **1** (4): 184-187.

Walkley, A. and Black, T. A. 1934. An experiment of the vegetative modification of chromic acid filteration method. *Soil Science,* **37**: 38-39.

www.ingramcontent.com/pod-product-compliance
Ingram Content Group UK Ltd.
Pitfield, Milton Keynes, MK11 3LW, UK
UKHW021533300726
14060UKWH00011B/509

9 789389 56994(